INDIAN AGENCY
NORTH DAKOTA
MINNESOTA
N
W
E
S
BIG STONE LAKE
Fourche
BLACK HILLS
SOUTH DAKOTA
James R.
Mississippi R.
Missouri R.
TRAIL
NEBRASKA
IOWA
YENNE
OGALALA
Platte R.
LINCOLN
ESBURG
OREGON TRAIL
ROCK CREEK
QUINCY
uth Platte R.
WESTERN
Kansas R.
INDEPENDENCE
KANSAS
ABILENE
COUNCIL GROVE
KANSAS CITY
ELLSWORTH
SEDALIA
GREAT BEND
NEWTON
OLIVER LOVING
CHISHOLM TRAIL
WICHITA
FORT SCOTT
MISSOURI
DODGE CITY
OLIVER LOVING
BAXTER SPRINGS
SEDALIA ROUTE
SANTA FE TRAIL
CAMP SUPPLY
ENID
TULSA
ADOBE WALLS
OKLAHOMA
FORT GIBSON
ARKANSAS
FORT SMITH
GOODNIGHT ROUTE
ANADARKO
Arkansas R.
CHISHOLM'S STORE
BOGGY DEPOT
FORT SUMNER
DOAN'S CROSSING
RED RIVER STATION
COLBERT'S FERRY
CHISUM'S BOSQUE GRAND RANCH
STAKED PLAINS
DECATOR
BOLIVAR
Red R.
DODGE CITY TRAIL
DALLAS
SHREVEPORT
TEXAS
FORT WORTH
PALO PINTO
NACOGDOCHES
WACO
ALEXANDRIA
HORSEHEAD ROUTE
SHREVEPORT TRAIL
LOUISIANA
Pecos R.
NEW ORLEANS
HORSEHEAD
AUSTIN
SAUNDERS
HOUSTON
W.D.H.
SAN ANTONIO
Rio Grande R.
CORPUS CHRISTI

Santa Clara County
LIBRARY

Renewals:
(800) 471-0991
www.santaclaracountylib.org

THE TRAMPLING HERD

THE STORY OF THE CATTLE RANGE IN AMERICA

THE
TRAMPLING HERD

PAUL I. WELLMAN

Illustrations by F. Miller

DOUBLEDAY & COMPANY, INC.

Garden City, New York, 1951

To that scholar and gentleman and high authority on the history of Western America, Dr. Frederick C. Narr.

Among the many who have with generosity helped in the collecting of material and the preparation of this book, the author acknowledges an especial indebtedness to W. A. Cochel, notable authority on the cattle industry, and Dr. Frederick C. Narr, able student of Western history, both of whom read critically the manuscript and made invaluable suggestions; also to Kirke Mechem, secretary of the Kansas State Historical Society, and Floyd Shoemaker, secretary of the Historical Society of Missouri, for placing the resources of their societies at his disposal; and to *The Kansas City Star* for the use of its invaluable archives on the cattle country.

FOREWORD

A prime difficulty in any attempt to tell the story of the cattle land of America lies in the inexplicitness of the borders both of territory and of time. Cattle are grown throughout the continent, and men handle them and, sometimes, profit from them. Yet to call Illinois or Virginia cow country would be a manifest ineptitude.

There is, however, a vast and somewhat indefinite area which, almost vaguely, attaches itself to the life and period we know as Western. Western does not apply to California or the Pacific Northwest. It applies to the territory between the Pacific coastal ranges and the general north and south line established by the Missouri River. It includes the whole of the great arid interior basin, the deserts, the plains, the brush country, the plateaus, and the mountains. Pretty well it defines the original habitat of the buffalo, of the prairie dog, of the coyote, of the antelope and of the lobo wolf.

And the Westerner? History indicates that the great population movements on this continent have been from East to West. The gold rush of '49 poured across interior America by way of the Santa Fé, Oregon, and Marcy trails. Later came the steady encroachment of settlement—working always from the East, gnawing bit by bit into the open range, eventually lapping over every bit of tillable soil not reserved arbitrarily by the government, and even turning with the plow the thin sod in areas which Nature had held only by a struggle and which, thus unbalanced, became desolate.

On the other hand the population movement which filled the cattle country has been largely neglected by history. It was from the South to the North. Two hun-

dred years were required by the East-West movement to
occupy that part of the continent east of the Mississippi.
Fifty years were needed for land emigration to finish pre-
empting all the arable land west of the Mississippi. But it
took only one decade and part of another for the cattlemen, streaming northward with their horn-spiked herds,
to fill the great interior of the continent. And today the
ways of living, the ideas, and even the talk of the range
country still are predominantly Southern.

Easterners are apt to look upon the West as raw and
new. It may be of value, therefore, to know that there
was a lively, well-grown cattle industry in the Southwest
long before the white man set his foot on Roanoke Island
or Plymouth Rock. Next to war, and possibly to mining,
the raising of cattle is the oldest industry of the white
man on the North American continent. It began within
two years of the start of the conquest of Cortés. It is
flourishing today as strongly as ever. It has, in the intervening centuries, contributed possibly the most distinctive
customs and mental attitudes of America, and certainly
some of the most stirring chapters of its history.

PAUL I. WELLMAN

Kansas City, Missouri.
September, 1939.

CONTENTS

BOOK ONE

HORNED IMMIGRANTS IN NEW SPAIN

BOOK TWO

HELL AND HIGH WATER

BOOK THREE

CAPITALS OF CATTLE LAND

BOOK FOUR

THE COWMAN TAKES OVER

BOOK FIVE

HIGH TIDE ON THE HIGH PLAINS

BOOK SIX

BLOOD ON THE SADDLE

THE TRAMPLING HERD

1. . . . . HORNED IMMIGRANTS IN NEW SPAIN

I

THE VENTURE OF DE VILLALOBOS

In the year of our Lord 1521 and the ninth of the reign of His Most Christian Majesty Charles V, a Spanish ship scudded before the fresh gulf breeze toward the low-lying misty blue cloud which was the coast of Mexico. It was one of the craft which, with increasing regularity, had been plying the Spanish Main between the islands of the Caribbean and the shores of New Spain for two years now . . . ever since Don Hernando Cortés set his iron heel and his still more iron grasp upon that rich land.

Most unfortunate it is that we have no description of this ship, for she was on an errand of destiny. Possibly she was a caravel—that being a common vessel of those waters and times—with a high narrow poop, a lateen sail abaft and square sails before, the cross of St. James and the striped flag of Spain snapping at the peak, and bulwarks gleaming with brass, for the Spaniard always has been a lover of the ornate, and never more so than in the period of the Great Conquest. Probably the persons who walked her decks differed little from other adventurers who had preceded them and would follow them to those misty shores. More than likely they were of the common ruck of fortune hunters bound for New Spain—soldiers, merchants, slave buyers, impoverished gentlemen, priests and tarry sailors. To all appearances the ship was just another ordinary Spanish craft bound for New Spain; that was all.

But a closer inspection would have revealed below the high poop, in the waist of the vessel, that which would have distinguished her from all her sister craft. Slipping and bracing themselves in their stanchions as the deck reeled with the tossing of the waves, was a little group of calves, their eyes starting in amazement and terror at this strange, shifting world upon which they had to scramble for a footing. The calves were consigned to New Spain. Theirs was a destiny sufficiently high, for they were to be the first of their kind to set cloven hoof on the mainland of the New World; they were to bring to the continent something which would change the course of history and remake the map of America—the blood strain that produced the longhorn of the West, and through him the cattle country.

It is not known with certainty how many calves were on the ship. One account says seven—six heifers and a young bull. It is, however, known that they were from Santo Domingo and that they were of the Andalusian breed introduced thirty years before into the West Indies by the Spaniards when they began exploiting those lovely islands.

A certain Gregorio de Villalobos was the shipper. Almost as little is known of him as of his cattle. That he was a Spaniard is beyond peradventure, with, as his name indicates, the blood of grandees. Whether he was short or tall, bearded or smooth-shaven, a soldier or a dilettante, we have no manner of discovering. But there is this to say for him: he, first of all his nation, saw the possibilities of a cattle industry in New Spain while every other Spaniard was crazily hunting for Aztec gold.

Two years before, in 1519, Cortés, after marching to Tenochtitlan, the capital of Anáhuac, told the Emperor Moctezuma, "I have a disease of the heart that only gold can cure." Most Spaniards suffered from the same ailment. It created in them a kind of madness which caused them to perform marvels of courage and cruelty. With the mer-est handful of armored swordsmen, Cortés overthrew a civilized native empire which could muster thousands of desperately brave warriors to every man of his. Not entirely by fighting, but in large measure through a superior technique in practical treachery, coupled with brutality so ruthless that it shocked and appalled even his fierce adversaries, the great conquistador achieved this miracle.

In it he was aided by the animosities of the Indians toward one another. Throughout the white man's history on this continent that same deep-seated enmity of tribe for tribe has played inevitably into the invader's hands.

Cortés found the Aztecs ruling a loose and somewhat rebellious confederation of tribes. About Tenochtitlan, the native capital, were nations never conquered by the Aztecs, and other nations which were unwilling vassals, aching for an opportunity to rise against their lords. For Cortés this was a sending of the saints. He played the Tlaxcalans and other disaffected tribes against the Aztecs. With reckless valor he threw his own steel-clad body and those of his men into the battles. He lied, a smile upon his bearded lips and a hand over his heart, with such smooth appearance of sincerity that the Indian leaders believed him—until it was too late. And always his seeking, and that of his followers, was gold.

But Gregorio de Villalobos saw in New Spain a different sort of opportunity. He may have been one of many Spaniards who observed the striking ecological similarities between the dry uplands of the New World and those of the cattle-raising provinces of Western Spain, but he was the first to put that similarity to advantageous use. While Cortés remained obsessed with his dream of treasure, Villalobos in 1521 brought his tiny seed herd from Santo Domingo.

It was a virile and adaptable cattle breed Villalobos introduced. Andalusian cattle, quick-footed and sharp-horned, were subject to wide mutations. For example, when turned loose upon the islands of the Caribbean, they speedily became wild as deer, as fleet almost, and alert either to fight or escape. Men made a living hunting these wild cattle, smoking their flesh for sale to the coasting ships, in rough smoke houses called *boucans* from an Indian word. Out of this term sprang the French *boucanier*, a generic title by which the hunters and driers of meat were known. After a time the hunters, whose lives were as savage as any heathen, began to prey upon ships which had the misfortune to be wrecked along the shores of the islands occupied by the *boucans*. Finding the profits excellent, they next banded together, eventually obtained ships of their own, and fared boldly forth into the gulf to capture haughty Spanish galleons bound for the treasure houses of the great king himself. So evolved, out of wild cow hunters, the famous buccaneers of history, progenitors of the pirates of evil memory. But that is another story and took place long after the landing of Gregorio de Villalobos' first little herd in New Spain.

Other than the wild cattle of the Caribbean islands, two historic breeds sprang from the Andalusian stock. One pro-

duced the fighting bulls of the Spanish ring—a breed of
peerless courage, quickness of foot, and deadly purpose,
the valor of which brings applause even from the sated
Spanish *aficionado*. The other was the longhorn of the
American plains.

Just where Villalobos established his first cattle ranch is
not of record. But his herd must have prospered. The Spanish cattle were small, dark and thrifty. Accustomed to the
barren Spanish hills they no doubt made good forage on
the coastal grasslands of Mexico. They increased. Other
Spaniards, observing the prosperity of this herd, imported
cattle also. Soon the numbers of animals grew through
increase and importation until herdsmen were required.
So the first cowhands of America appeared.

After the treachery of Cortés at last was discovered
and the Aztecs rallied under their heroic leader Cuahtemoc
in a hopeless, bloody resistance in beautiful Tenochtitlan,
the Spaniards, when they wearied at last of slaughter, took
many captives. Of these the worthless were slaughtered
out of hand, but the strong, healthy, and tractable, Cortés
caused to be branded on the cheek with a G for *guerra*
(war) and sold as slaves. Many wealthy Spaniards in
Mexico possessed some of these slaves. It can be speculated
almost with certainty that Villalobos had a herdsman to
care for the first tiny seed herd of cattle in America who
wore an angry red G scarred on his brown cheek. If
so, by a curious twist, the first cowboy in this hemisphere
bore a brand before the first cow.

Within a decade cattle had a strong foothold on the
new continent. Yet even with the passage of many additional years their numbers remained scanty compared to
the immense herds of a native representative of the bovine
family—the American bison. Cortés and his men saw one
of those strange creatures in the menagerie of Moctezuma
when first they entered Tenochtitlan. Afterwards they

described it with awe as "the greatest Rarity . . . the Mexican bull; a wonderful composition of divers animals. It has crooked Shoulders, with a Bunch on its Back like a Camel; its Flanks dry, its Tail large, and its Neck cover'd with Hair like a Lion. It is cloven footed, its Head armed like that of a Bull, which it resembles in Fierceness, with no less strength and Agility."

Could Cortés and his men have known the whole truth about the "Mexican bull," they would have been astonished beyond any possible words. The "crook-backed cows" in that day grazed in such myriads on the plains north of Mexico that their numbers were beyond computation. Of the great central area of the continent, they were the lords. Yet destiny is strange; the tiny herd of calves, on the day it was landed at the orders of Villalobos on the coast of New Spain, sounded the death knell of the millions to the north. To provide pasturage for the descendants of these and other cattle imported across the sea in increasing number, the white man one day would obliterate the buffalo herds in the greatest slaughter of living creatures known to history.

For the present and for some three centuries to come, however, no such cataclysmic event occurred. Slowly at first, then more rapidly, the Spanish spread their cattle across Mexico, so that by 1540 when the visionary captain, Don Francisco Vasquez de Coronado, made his long and disappointing march up into the high plains of what is now the central United States, there were cattle ranches on both the west and east coasts of New Spain. In his golden mail, with white plumes floating from his helmet, young Don Francisco rode out of Compostela on the west coast of Mexico, February 22, 1540, at the head of a large armed force seeking the fabled seven cities of Cibola. In the rear of his army were driven large droves of slaughter animals, including sheep, hogs—and cattle.

The adventures of those animals, particularly the cattle, are somewhat dim in the accounts of the expedition. Shortly

after the army left Compostela several days were lost at
the Rio Centizpac, because the "cattle" had to be taken
across. It is probable that the translation of the Spanish
word should be broader—livestock. The beef herd contained perhaps five hundred head. Many animals of all
kinds were killed or lost by the time Coronado reached
Culiacan, on the river of that name in Sinaloa, which at
the time was the last northern outpost of Spanish civilization on Mexico's Pacific coast. The accounts further say
that when the army reached Culiacan the livestock was
wearing down its hoofs, and pack horses and even mules
had fallen and been killed in the narrow mountain paths.
There is some reason to think that perhaps the only livestock which actually reached Cibola was horses and sheep.
Paul Jones, one of the highest modern authorities on Coronado, is of that opinion. There was no need for meat or
hides of domestic cattle in the later stages of Coronado's
march, because the amazing herds of buffalo were encountered and found easy to kill. Jaramillo, one of the
chroniclers of the expedition, wrote: "There is profit in
the cattle ready to hand."

So many of Coronado's cattle were abandoned in Sinaloa
that twenty-five years later Francisco de Ibarra discovered
thousands of their descendants running wild in the province. Within the next few decades great ranches sprang up
in that territory, one of them possessing so many cattle that
it is reported to have branded thirty thousand calves in a
year.

Yet although all of Coronado's cattle may have been
slaughtered or lost when Cibola was reached in central
New Mexico, a few probably still were with him when
the expedition first crossed the present border of the
United States, entering Arizona almost directly south of a
native village called Chichilticalli, where now stands Fort
Grant. On more than one count that was a historic occasion. Not the least important aspect of it was that those
slowly ambling cattle were the first of their kind to enter

what is now the United States, and that the journey they were making partook of many of the characteristics of a history-making institution, the trail drive, that was destined to remake the American West.

By the time Coronado left Tiguex, on the Rio Grande River, April 23, 1541, it is recorded that the slaughter animals were nearly all gone. And when at last the young adventurer, disillusioned and bitter at his failure to find the cities of gold, returned to Mexico in 1542, the record says "no slaughter animals were left." We know, therefore, that if any of Coronado's cattle reached the present territory of the United States they became beef, unless a few of them were lost.

And here arises a fascinating speculation. It has long been a theory that some of Coronado's horses and cattle escaped and survived to perpetuate their breeds, thus beginning the wild herds of mustangs and longhorns on the Southwestern plains. Particularly is it possible that this was true of the horses. The Spanish cavaliers liked stallions for war steeds, and there were mares in the cavalcade which accompanied Coronado. It is known that horses were lost, and it is at least a strong possibility that some of them lived and bred on the great grasslands. The belief that these escaped horses of Coronado were the ancestors of all the thousands of wild horses which roamed the western plains when the white men began to cross them in the early 1800's is, however, scarcely justified. By then the blood of the Coronado horses, if any survived, had long been augmented by estrays from the Spanish ranches in New Mexico and Texas which had slowly spread northward.

There is even less likelihood that the cattle perpetuated themselves in any permanence. If only because so many were lost in Sinaloa, the few remaining when New Mexico was reached would have been watched with double care. Even if a handful did escape from time to time, they probably were too scattered to do much propagating.

But if cattle, in their first entry into the present United States failed to survive for long, others soon were headed back on a far more permanent basis. The Spanish were natural cattlemen. The first white man to cross the plains of Texas bore the significant name of Cabeza de Vaca (Head of a Cow). His descendants still run cattle in Arizona. Cortés himself named his *hacienda* in Cuba, where he lived before his Mexican expedition, Cuernavaca (Cow's Horn). Favorable conditions combined with the natural cattle-raising instinct of the Spaniards to increase steadily the number of ranches in Mexico. Two years before Coronado's expedition the government decreed the formation of a stock raisers' association, and a law was passed against the castration of bull calves for fear the herds would not propagate as rapidly as desired. Before his death Cortés himself once more became a cattleman, and he is credited with first using the branding iron in America. Already he had used the brand, but on human beings. Now he used it also on his cattle. That ancient brand of Cortés, three Christian crosses, was the forerunner of the whole remarkable wild heraldry which still exists today in the range country.

With settlement, the cattle ranches gradually spread north along the two seaboards of Mexico. When Coronado marched, the Spanish frontier in New Spain extended in a rough semicircle from Culiacan on the Gulf of California, to Panuco (Tampico) on the Gulf of Mexico. But led by mining interests exploring the mineral resources of the country, by zealous missionaries of the Catholic church, by slave hunters, and by wealthy cattlemen seeking new range, the line of settlement pushed farther and farther north. Particularly under Gaspar de Zuniga y Acevedo, Count of Monterey, the ninth viceroy of New Spain, was interior expansion rapid and vital. In 1581 the Spaniards once more marched into the Pueblo Indian country of New Mexico and again men talked of locating the fabulous kingdom of Gran Quivira and the equally mythical Strait of Anian. To attempt to find these Cantaño de Sosa led an expedition in 1590, and Antonio Gutierrez de Humana and Francisco Leiva de Bonilla in 1593.

The real Spanish foothold in the American Southwest, however, was obtained in 1598 when Juan de Oñate established a temporary colony in the Rio Grande valley near El Paso and sent his lieutenants to explore the Colorado River to the Gulf of California on the one side, and the plains as far as Kansas on the other.

Even earlier than this, however, the Spanish thrust had gone north in another direction. Luis de Carabajal secured, in 1579, a grant to Nuevo León from Martin Enriquez, the fourth viceroy. Using as a base the old settlement of Panuco, Carabajal marched west and in 1583 laid the foundations of the town of León, now named on the map as Cerralvo, in the Rio Grande valley not far below the present border of Texas. At about the same time the settlements of San Luis, near present Monterrey, and Nuevo Almadén, near Monclova, were established. The ugly motive behind this particular conquest was the hunt for slaves. Several hundred Spaniards joined the expedition, brutally rounding up the terrified Indians and selling them to ranches, mines and plantations.

Thus were the Spanish settlements established, and by 1590 the frontier stretched from Cerralvo on the Rio Grande, by way of Saltillo and San Bartolomé, to San Felipe at the mouth of the Sinaloa River. It must have been very soon after the town of Cerralvo began to grow near the Rio Grande that the first cattle herds were crossed north of the river to pasture. At least one Texas historian estimates that the first herds were grazed in what is now the Lone Star State by the fall of 1583.

A vital part was played by the missions in the northward migration of the cattle herds. The church's way was thorough. As they entered a new territory, the *padres* sought to teach the natives the principles of farming and livestock culture, bringing their own cattle along with them for this purpose. At first the process of education was slow, but gradually, with the painstaking completeness which characterizes all the functions of the church of St. Peter, the lesson became ingrained. From the Spanish priest the Mexican Indian received his first lessons as a *vaquero*. In turn he taught the American cowboy almost everything he was to know about handling cattle.

To those who think of the cattle industry of the West as something recent and raw, consideration of the year 1580, the beginning of the decade in which cattle first were introduced north of the Rio Grande, should prove illuminating. In 1580 Philip II sat on the throne of Spain and Queen Elizabeth on that of England. The Spanish Inquisition was at its height, and the Holy Office, with its superiors and familiars, held dread control not only over Spain but over Spanish America, and many parts of Europe. Not for eight years yet would Philip make his climactic effort to subdue England and send his great Armada to destruction in the English Channel.

Englishmen, in 1580, were still talking about the recent loss of Calais to the French. Mary Queen of Scots lay in prison and was to lose her charming head within seven years. A certain Will Shakespeare was whiling away his time at Stratford-on-Avon, and perhaps doing a little deer poaching on the side—a habit which was to cause him to depart suddenly for London five years later and to eventual immortality.

Not for twenty-seven years would the first permanent British colony be established on the mainland of America at Jamestown, and the English nonconformists would not land at Plymouth for forty years. In 1580 the Southwest would have been justified in considering the Atlantic coast a barbarous wilderness. The cattle industry of the West had already begun a sturdy development, while the English were still without their dream of empire.

II

LONELY WATCHERS BY LITTLE WATERS

In the immensity of the desert and the plains, from the Rio Grande's lower valley clear over into New Mexico, and up the Santa Cruz, San Pedro, and Sonoita valleys of Arizona, tiny herds of cattle began to appear, seeming merely to accentuate the vast emptiness. The buffalo range was remotely to the north of those first little ranches. Indians were in this border land, but on the plains the predatory tribes lived on the edges of the bison herds. Because the horse was yet to come into general use among the Indians, the savages traveled perforce on foot. Their mobility was thus limited and this prevented them in those first days of the cattle country, while the ranchmen were obtaining a foothold, from becoming the raiding scourge of later years.

It was well that at first the Indian menace was not of a type to exterminate the first settlements—skirmishes with the aborigines were common enough as it was—because those early stockmen were embarked on an adventure perilous enough at best. Deeper down in Mexico and particularly along the littoral of both oceans, the Spanish swords had brought about a condition fairly stable, with

24

the natives for the most part subdued and opportunity already existing for peaceful development. In those fortunate areas were established the great *estancias*, inland feudal empires, with their wide-spreading *haciendas* and their *hidalgos*, virtual masters of their peons and peasants and wielding the authority of the high justice, the middle and the low, with the royal prerogative of taxation, often, to supplement the wealth they gained from their rapidly growing estates. They lived on a lavish scale, those old grandees, so much so that when the bishop of the province came to call upon one of them, the prelate found a walk of pure silver constructed so that the episcopal robes might not sweep the dust between the coach and the palace.

North of this section, however, the frontier was an implacable wilderness. Not only their fortunes but their lives were taken into their hands by the venturesome Spanish cattlemen who first crossed the Rio Grande. The spread of the herds consequently was slow.

Along the lower Rio Grande the country was almost tropical. Live oak groves stood draped in funereal Spanish moss. Mesquite and chaparral invested the landscape with what seemed an impenetrable jungle, in which lurked puma and jaguar, with the fierce little musk hogs known as peccaries or *javalinas*, many poisonous reptiles and much game, such as deer, antelope, and wild turkey. Westward the vegetation was progressively less dense until finally it scattered out and eventually disappeared almost entirely in the sands of the deserts. In the southern areas of New Mexico the grama grass was nutritious, curling and curing in the summer so that it formed a first-class hay on the root, which lasted through the winter. Wherever the *padres* or soldiers could erect fortifications sufficiently strong against the depredations of the rapacious tribes, there were cattle, together with horses, sheep and goats.

In the desert country the grim exigencies of constant lurking war constricted the spread of the herds. The Pueblo revolt of 1680-1682 wiped out the settlements of New

Mexico, and when De Vargas reconquered that territory, the new settlers had to begin all over again with cattle and other livestock brought anew from the southern provinces. It was in southern Texas, therefore, that the Spanish cattleman established himself first and most firmly inside the present borders of the United States. New Mexico was, and remained, more a sheep than a cattle country until after the War between the States. And then it was Texas trail drivers and Texas longhorns which gave it a sudden new life as a cow country.

Between the first primitive Spanish ranches in the Southwest and the cattle ranch of today or even of half a century ago there is little to compare. The life of a cowboy of the '80's can scarcely be called the existence of a dilettante, but compared with the way the first cattle herders lived a century and a half before, he might almost be said to have had a luxurious career. We have seen how Aztec slaves with the letter G branded on their cheeks watched the herds of the Spanish *hidalgos*. Long after the slave trade ceased the Indian tradition clung to the cattle range. Later the Mexican, in whose veins ran white as well as Indian blood, and who combined the indolence of the red man with the daring of the Spaniard and perhaps some of the Spaniard's treachery, superseded the Indian.

The early Mexican herder lived in surroundings as savage as the wildest Indian. Lewis Garrard, who visited a remote cattle outpost in New Mexico before the United States conquered that territory, left a vivid vignette of life there. Many years had passed since the Spanish first pushed their herds north of the Rio Grande, but conditions were at least as primitive when Garrard observed them as in those earlier times, and we may believe that the Mexican *vaqueros* he saw were fairly typical of their predecessors in the desert and mesquite.

No ranch house sheltered those half-savage cattle guardians. They lived in crude lean-to huts, built where the essentials of a camp—wood, water, and shelter from the coldest winds—could be found. Two forked poles planted upright in the ground, with another pole laid across the forks, formed the simple framework of such a shelter. Against the transverse pole were leaned smaller poles, their butt ends on the ground, forming a sloping surface over which a covering of rawhides managed to shed the rain, to a degree at least. Upon bedding of hay were spread robes and blankets to form a lounging and sleeping place for the men, and also a floor. No side walls; a shelter at the back was considered sufficient by those defiers of the elements, and besides, there was a fire in front, while the rear of the lean-to was turned to the prevailing winds. Such a dwelling would, of course, have been unthinkable farther north, in Wyoming or even Colorado. But in Mexico, southern Texas, and New Mexico—wherever winters were mild—it served.

For months on end the food of the early Mexican *vaquero* consisted of a monotonous diet of *atole* mush, supplemented by such game as he was fortunate enough to kill. Even in hunting he was vastly handicapped as compared to his white successor, for he possessed no firearms. Instead he depended on a bow and arrow, in every way as barbaric as those of the wildest Comanche or Apache. And on his *reata*.

Yet already in this crude early representative, the way of life of the American cowboy had its beginning. For one thing the *vaquero* never walked if he could ride—a prejudice against pedestrianism which he handed down with full force. He prided himself, moreover, on being able to ride anything with four legs and hair. Sometimes, to prove his boast, he displayed his skill on the back of a bouncing and bawling longhorn steer, and there was no pony in his remuda, no matter how much of an outlaw, which had not been tried out so that its every secret of speed, stamina, and performance was known.

Usually the early *vaqueros* went barefoot, their slender

pay and the mild weather combining to eliminate boots. When they shod themselves at all it was with straw or rawhide sandals. Yet every *vaquero* managed to scrape up enough money to buy, or in some other manner obtained, a pair of murderous Spanish spurs with great twirling rowels, which he strapped on his bare heels and used to rake the sides of whichever miserable beast he happened to bestride. He preferred, moreover, to wear those spurs even when he was not riding, as a sort of badge of his calling. And in this, too, he was imitated by the American cowboy of later years.

There was in this remote beginning already the great virtuosity of the rope, that universal skill which cattlemen share in some form throughout the continental range country. Cattle can always be driven, but to be controlled, branded, and otherwise handled, they must be caught. And it is not always possible to find convenient corrals into which to drive them—especially in a wild, uninhabited country. Therefore the *gaucho* of the South American pampas swings a *bolas*—a rope with weights at its ends, which whirls through the air and entangles its quarry. But the *vaquero* of North America, and the cowboy who followed him, preferred the lariat. From the Mexican the American cowpuncher learned the use of the rope, a skill which had been evolved in the generations during which the Mexican descended, first from the cheek-branded slave, and then through long service as a wild herdsman in the lonely lean-to shacks.

The *vaquero* lived with his lariat. In his hands it became an inspired extension of his arms. From the saddle it enabled him to fasten on and throw the biggest steer, to noose and choke down the wild stallion, even to capture game as fleet as deer or as ferocious as the grizzly bear— the last feat depending, to be sure, upon whether there were other *vaqueros* handy with ropes to help in catching the foaming beast, stretching him out, a hind leg this way, a forepaw that, and his neck in still another noose, encircled with the strangling cord until the breath ceased to come, the foam dripped from his feebly gnashing jaws, and the wicked green lights in his little pig eyes glazed and dulled in death.

So lived the early *vaqueros*—they cannot be called cowboys, because, except for the mutual relationship to the horned herd, the two had little in common. Still, those early, savage Mexican herdsmen established the traditions of loyalty to the outfit and devotion to the cattle, as well as reckless disregard for personal danger in the line of duty, a tradition which, incidentally, continues today to imbue the cattle country. For months at a time the *vaqueros* lived far away from their families, with no eye to watch their doings, yet they prided themselves in caring for the herds of their employers more carefully than if they had been their own. The blasting heat of the summers they abided, and the cold rains and winds of winters, when they crouched about their smoking fires or lay beneath the dripping eaves of their huts. Sometimes they almost starved, but it was a last resort when they killed a steer, for this was the property of the outfit and therefore inviolate.

Always, too, in those early days, the *vaqueros* lived in imminent danger of death at the hands of the Indians, for the horse soon became the property of the tribes and there grew up a regular system of raiding along the frontier. In their isolated outposts the Mexican *vaqueros* knew and feared the savages from the Staked Plains or Apacheria. Yet with a devotion pathetic and admirable they remained with their cattle. Nobody knows how many almost naked, almost unarmed, humble heroes of the herds died in the rush of whooping marauders out of the wilderness.

In this early period of the cattle industry another principle was established—the principle of water rights. The loneliness of the *vaquero's* existence was due to the nature of the country in which he lived. Grazing land there was in abundance—more than anyone could use in that day. But if grass was plentiful, water, in sufficient quantity to support any large number of cattle, was scarce. Along the rivers water could be found, and the riparian rights were soon pre-empted. Out from the large rivers every little streamlet, *laguna*, and desert tank was prized like a precious jewel. Possession of water was tantamount to possession of the range surrounding it.

Many of the first *rancheros* took up land in the new regions by virtue of royal land grants from the king of Spain—parchments entitling them to possess so many leagues of territory, which they usually located for themselves. Water was what determined this location—the best water the owner of the royal grant could find. Thus it came about that Spanish land grants usually were quite definite in reference to the *rio*, the living spring, or the lake upon which they were situated, but otherwise were often vague in their descriptions—the boundaries frequently beginning at a certain stone, running to a certain tree, and so on, likely to vary with the passage of years, but unimportant in that day, since the use of the range by the law of common consent was determined by its accessibility to water.

As late as the '70's John Iliff, one of the great figures in the northern cattle range, insured himself immense grazing grounds, when the first change began from free range to titled property, by taking up frontage along the Platte River. He filed on a homestead for himself on the river bank, and had his friends and employees also homestead, always along the river. As fast as the claims were proved up, he acquired the titles, either by purchase or by an understanding with the prover. In the end he owned the bank of the Platte for upwards of thirty-five miles, and the grass behind his water front was his as far as the cattle would range.

And that was a long distance, for the old Texas longhorn had a prodigious capacity to range. It was not uncommon for a longhorn steer or dry cow to walk fifteen miles out from water to where the grazing was good in times of drouth—and walk back every other day to drink. An almost camel-like capacity to live for long periods without water seemed to be developed by them. On the Horsehead route cattle traveled continuously three or four days and nights without a drop of moisture, and in the winters longhorns existed in the brush country of Texas for months at a time, with apparently no water except such as could be obtained by chewing the pulpy stalks of prickly pear and other cactus, to which they became so accustomed that the bristling spines in their lips and tongues seemed to bother them little. With a herd of Texas longhorns a man who possessed a good water hole could count on his cattle using three or four hundred square miles of grazing territory, extending in all directions.

Early in the West a tacit agreement protected water rights, but there were always violations of the code. The first *vaqueros* probably were given the responsibility, among other duties, of watching their own water holes to prevent the cattle of other *ranchos* from using them. This we know because from earliest times has come down the line-riding system by which *vaqueros* or cowboys, riding an imaginary line which is the understood limit of their range, daily made a habit of "throwing back" cattle belonging to their neighbors, thus preventing a drift. Not a few fierce little range wars were fought between line riders over a creek or a series of water holes. Many a line house, particularly in the Southwest, was built like a fort, not only as a defense against the loping Apaches, but if need be against interloping neighbors.

Water rights, based in many cases on mere squatter claims, created ranch sites; and since water was so scattered in most of the ranch country, the ranches also were scattered, and their men, the primitive *vaqueros* of the bow and arrow, the *lazo*, and the huge Spanish spurs strapped on bare brown heels, were scattered too. In those isolated outposts as well as in the larger interior ranches were developed the methods of cattle handling which in the essential remain in general use today. How indelibly the Spanish-Mexican influence stamped itself on the cattle industry is seen in the range language.

Words like corral, tapaderos, remuda, segundo, broncho, jinete (broncho buster), orejana (an unbranded steer), frijoles, hombre, loco, are almost pure Spanish. Other words have been more or less twisted in transition. There is little difference between lasso and *lazo*, rodeo and *rodear*, doby and *adobe*, pinto and *pinta*, lariat and *la reata*, or ranch and *rancho*. But some of the transitions have gone even farther. The cowboyism, cavvy yard, for instance, is from the chaste Spanish *caballado*. One of the oddest evolutions is the name for the act of wrapping the end of a lariat two or three times about the saddle horn in roping an animal, in order to throw it. The Spanish phrase for this was *da la vuelta*. To the cowpuncher this has degenerated to an approximation in sound if not in written appearance—dolly welter—or simply dolly.

III

THE FORLORN LONGHORN

No odder, weirder cattle specimen ever existed than the
Texas longhorn. He was the clown of his family, a carica-
ture of his kind, in which every peculiarity was accentu-
ated to degrees which made of him almost a walking car-
toon. His legs were too long and too bony. His tail was
too long, and his body was thin and so long that often he
was sway-backed. In color the longhorn was unpredictable,
sometimes brown, sometimes dun, sometimes red, sometimes black, occasionally yellow, and often an indiscrimi-
nate combination of all colors. More than all else his
horns were outrageous and unbelievable. Out and up, on
either side of his head they curved, sometimes almost
corkscrewing, making a prodigious sweep, polished white
or blue and tipped at the end with points sharp as
stilettos. Looking at those mammoth horns the spectator
was often impelled to wonder how the scrawny neck of
the creature could support such magnificent adornments.
The longhorn steer reached his full growth when he was
about ten years old, but not so his horns. Year after year
they continued to lengthen and spread as long as the animal lived. Old steers often carried horns measuring seven

feet from tip to tip, and there is a record of eight feet, nine inches. The bases of these mighty horns became wrinkled and cracked, giving the appearance of lichens clinging to a rock. That was what gave rise to the descriptive term "mossy horns" referring to mature cattle in the trail-driving era.

Naturally sad-eyed and hang-dog in his appearance, the longhorn became even more so when man added to his natural homeliness. John Chisum at one time had between seventy thousand and one hundred thousand longhorns ranging up and down the Pecos River under his brand, the Long Rail and Jingle Bob. The Long Rail was simply a line burned from the shoulder of the steer along his side to the point where the hair ended on the flank. It was a type of "barnside" branding very popular in the early days before hides began to attain a larger value, because it was almost impossible to blot. Ace Jenkins, for example, put a huge A on the shoulder, a C on the ribs, and an E on the flank of his stock. That brand could not be altered. Later, when the demand for cow hides grew heavy, brands became smaller since they were found to damage the hides.

To return to Chisum's cattle. The Long Rail was unbeautiful in itself, but it was the Jingle Bob which really made walking freaks out of the steers that wore it. It was the result of knife work on the ears. The ear was split in such a way that part of it hung down flapping while the rest of it stood erect as Nature ordained. So distinctive was it that Chisum's cattle were known as "Jingle Bob cattle," instead of being called by the brand name as was the usual custom. Chisum was proud of the fame of his mark. It required skill to cut the ears exactly right, so he assigned the work to a few of his men who became specialists at it. When properly Jingle Bobbed, a steer had an appearance both ludicrous and repellent. With his absurd horns, six or more feet long, curved like scimitars and sharp as bayonets; his stupid, half-frightened, yet half-ferocious countenance; his lean, sway-backed body and long bony legs—to have four ears, two standing erect and two hanging down like those of a hound, made him a very specter of an animal, fit to frighten or amaze.

For all his idiosyncrasies in appearance, whether natural or man-created, the longhorn steer, however, was a worthy product of his race and his environment. As has been stated, the first cattle in America were of Andalusian stock. With this blood were mingled many other strains all of which went into the creation of the longhorn. Cattle in the Southwest were introduced to a new and vast wilderness. Often they were turned loose to shift for themselves, spreading always outward and northward from the fringes of the slowly advancing Spanish settlements. In that wilderness they found wild animals to combat and so developed characteristics which enabled them either to flee or fight as the need or justification arose. Traits which became noticeable in the later trail-driving days—the propensity to go into a blind panic and stampede over nothing, the savage and sullen temper, the terrific endurance— all were protective and were developed through evolution as were the fabulous horns.

In the mesquite country of Texas the Spanish cattle grew as wild as the deer, antelope, and *javalinas*. When the first Americans came into Texas they found a breed of little, black, sharp-horned cattle in the canebrakes and the brush. These became known as "mustang" cattle, or sometimes as *cimarrones* (wild things). They were not true longhorns, but a strain of outlaws, fierce and cunning brutes, as wild and dangerous as the black buffalo of South Africa, many of whose traits they seemed to possess. Mustang cattle, when captured, could hardly be tamed, and displayed the greatest disinclination to cross-breed with other stock. Even when captured as young calves they did not become domesticated and it was said to be easier to capture and tame mustang horses than cattle of the same type. Early settlers complained, moreover, that the wild cattle lured away their domestic cattle in much the manner that a wild stallion will toll off tame mares on the range.

Known to the early settlers as "black cattle," they were usually of that color, although there were variations among them, including reds, browns and brindles. Their horns were polished, needle-sharp and pointing forward—killing weapons, unlike the horns of the longhorns which spread widely and point up. Frequently a line of white or yellow went down the back of one of these animals, similar to the line-back seen on many buckskin mustang ponies. These cattle were the direct descendants of the Andalusian stock, and from the time that the first colonists from the United States under Stephen A. Austin began to enter Texas in 1821, the *cimarrones* were considered wild game, just like deer or bison. Colonel Richard I. Dodge, writing of different types of game in the Southwest, said:

I should be doing an injustice to a cousin-german of the buffalo did I fail to mention as game the wild cattle of Texas . . . animals miscalled tame, fifty times more dangerous to footman than the fiercest buffalo.

Instances abound in which wild bulls of the black cattle breed attacked travelers and even hunters, killing some and forcing others to climb trees for safety. One writer has left the record that it was dangerous to camp on the east side of the Little Brazos River in Texas "on account of the cattle coming in for water, the night being the only time they drink." This was the considered opinion of Buck Barry, one of the most famous of the old Indian fighters. He did not fear the Comanches, but he steered clear of wild cattle, even when they were only coming down to the river at night to drink.

On one occasion a mustang bull attacked single-handed —or rather single-headed—an entire American army on the march and threw it into confusion. It was the force of General Zachary Taylor, headed in 1846 from Corpus Christi toward the Rio Grande to invade Mexico, to which this humiliating experience occurred. A soldier marching on the flank of one of Taylor's columns fired at a wild bull which instantly charged, his tail up, his twin rapier-like horns down, and his nostrils snorting rage. The soldier who had shown bad marksmanship fled into the column for safety, but the bull thundered right after him. Full into the marching men he rushed, "scattering regiments like chaff," according to Colonel Dodge's account, with the troops falling over one another, files breaking into undignified rout, officers swearing and unable to do anything to halt the debacle. So tangled were men, weapons and the animal, that nobody dared fire for fear of wounding a comrade, and the bull at last galloped away, unharmed. Fortunately in this instance nobody was injured, but that bull possessed the distinction of having, alone, thrown Zachary Taylor's forces into confusion which the whole Mexican array was unable to do later at Monterrey.

Even more amazing is the account of an actual pitched battle fought between a herd of bulls and the Mormon Battalion, commanded by Colonel Philip St. George Cooke, on its way from Santa Fe to California during the same war with Mexico. The country through which the column was marching at the time, the San Pedro valley of Arizona, had been devastated by Mangas Coloradas' Apaches and there were no human habitations left in it. The soldiers, however, noticed many cattle trails going into the brush.

Suddenly, as Cooke later reported, a veritable squadron of bulls charged suddenly at the battalion as it passed through a wide grassy canyon. Seeing the beasts coming, the colonel ordered his men to load their muskets and repel the attack. Then some of the brutes plunged out of tall grass right by the path down which the command was marching. One bull caught a soldier with a horn through his thigh and threw him backward over its body, then rushed on a team of mules, killing the first and tearing out the entrails of the second. A huge, coal-black bull was slain by a corporal just before it reached Cooke himself, and the colonel wrote, "It fell headlong, almost at our feet." Several men were injured, some fatally, in this strangest of all battles fought under the United States flag. Some of the beasts also were killed and the rest finally withdrew, but it would be hard to claim a victory for the American arms in view of the destruction they left behind.

The wild cattle were far more savage than the Mexican cattle which were kept at the *ranchos* and remained somewhat domesticated. To the latter were added cattle brought to Texas by Austin's "first three hundred" settlers—bigger-boned, more docile creatures from Louisiana, Tennessee, Arkansas and Missouri, the Missouri strain being in particular famous as producers of yoke oxen. As other settlers followed the first group they also brought in their herds, American cattle which had descended from animals introduced at Jamestown, Plymouth and elsewhere on the Atlantic coast long after the ancestors of the Mexican cattle were thoroughly acclimated to the arid Southwestern country. For the first time the strains from the East and the South commingled, and the new blood graded up the original Spanish cattle, giving them added size, heavier bone, better flesh, and tremendous horns.

Because of the scarcity of water, the longhorns developed great powers of endurance and with that the ability to withstand thirst, heat, cold and hunger, a goat-like agility in climbing, and the quality of being able to live off almost anything. Alfred Henry Lewis tells of making friends with a young longhorn bull by giving him a strip of bacon rind. The bull, which had come up full of truculence, forgot all about his ill humor in the ecstasy of chewing on the salty rind. A ragged shirt, a straw hat, a pair of drawers—all were delicacies to a longhorn steer, which seemed to possess a belly as accommodating as either a goat or burro.

It was the vitality of the animal, however, which was its

most amazing characteristic. J. Frank Dobie has a good yarn of an old-timer who bought fifty high-grade "American" bulls, and turned them loose with his native cattle.

"And they all died," the old-timer added.

"Texas fever?" asked Dobie, remembering the cattle tick.

"No," replied the other, "the native cattle simply walked them fine-haired, short-legged bulls to death."

It was because of the underlying truth of that anecdote that the practice of shoeing blue-ribbon bulls has long been in effect in some parts of the Southwest. Otherwise the soft hoofs of the prize stock became so sore through trying to keep up with the native cattle, that eventually the animals refused to walk, hung around the water holes, and starved for want of grass.

Surly and morose the longhorn undoubtedly was, but he lacked the killing instinct of the black mustang cattle. Frequently a cow with a calf would go "on the prod," as the cowboys said, and fight like a fiend if she thought her offspring was being disturbed. Occasionally, also, bulls would become ugly. But most longhorns were controlled easily by a mounted man, and they were far more apt to run from than to charge a man on foot. There are, in point of fact, numerous instances when the sudden appearance of a pedestrian caused a disastrous stampede of a whole herd of longhorns. The beasts were accustomed to man's presence as an integral part of the horse-and-man combination, but the sight of him apart from his quadrupedal associate sent them into a crazy, drooling panic.

So developed the longhorn. In Texas he found an ideal clime for his evolution. During the long years when the range was wide open and free, the longhorn so spread and multiplied, that when the northern grazing lands at last opened, Texas was able to stock the entire West from her own huge reservoir of cattle supply.

IV

CENTAURS WITH RED SKINS

On foot the plains Indian was a furtive and rather pitiable figure—prowling and bushwhacking, trying to still-
hunt for game, often happy to find an abandoned buffalo
carcass so that he could fill his belly with carrion after
other resources had failed. But when he obtained the horse
he became a different being utterly. At once, overnight
almost, he burgeoned into a bold and tireless raider, a constant wanderer, a well-fed dweller in what to him was an
excellent habitation—the teepee—an exponent of color in
costume, furnishings, and personal adornment, a never-
wearied seeker after renown and a swashbuckling boaster
concerning his own feats. And he became also the greatest
hazard and menace to the frontier cattleman.

There were Indians in Texas when the first cattle herds
pushed north of the Rio Grande, but they were not horse
Indians. Caddoes, Orquisacos, and Nacogdoches, they were
fishermen and hunters of small game, almost timidly inclined to friendship with the palefaced newcomers. The
Caranchuas were more warlike. But it was not until the

40

early ranchers encountered the Comanches, that numerous, lethal, wily race, on the interior plains of Texas, that they learned what Indian trouble really was.

At what era the Comanches obtained their first horses is not known, but they rode as if born in the saddle. To them the horse was a familiar and essential auxiliary to most of their lives' activities. Possibly their first ponies were obtained from other tribes having closer contact with the white man. Possibly the wild mustang herds, which by 1600 were wandering on the Southern plains, supplied them. At any rate the Comanches, at their first introduction into history, already were accounted horsemen almost without peers, whose ability to ride was so legendary that the expression "he rides like a Comanche" was throughout the West the highest praise equestrianism could achieve.

The Comanche had little of the sentimental love of the Arab for his mount. There was, in point of fact, little room for sentiment of any kind in him—he was too busy with war. To him the horse was merely a means to an end. It carried him on his journeys, to war and to the hunt. It lifted him to a rank equal with the swiftest running creatures. It contracted distances and thus enlarged his horizons. For these reasons the Comanche prized the horse without loving it, and built his life economy about it.

It is not to be wondered, therefore, that the Rio Grande settlements early grew to know well and unfavorably the Comanche. North in the Powder River country of Wyoming the Sioux were expert horse thieves. In the valley of the Upper Missouri in Montana dwelt the Blackfeet and they even exceeded the Sioux in the purloining of other people's steeds. The Cheyennes, Kiowas, Pawnees, Crows, and Apaches each were bad after his kind. But compared to the Comanches all these were tyros and innocents in horse larceny. To the Comanches the remudas of the Rio Grande frontier were a constant and irresistible lure. Here were horses, many horses, ready to hand. All that was needful was to ride down through the thinly populated and scattered white settlements, round up whole herds of ponies, and be off. Escape was easy, because the booty furnished its own locomotion which was as rapid as any pursuit.

True, the Comanche, when he was raiding, was most catholic in his tastes. He did not scorn to drive off cattle with the horses, to carry away weapons, clothing, or anything else he fancied, to kidnap women and children, and to pick up a scalp here and there as opportunity offered. The horse raiding of the Comanches was the genesis of the trouble between the Texas cattlemen and the Indians. Later, with the connivance of the Comancheros—that strange gypsy trading population of the Staked Plains— stolen cattle acquired a high market value and longhorn steers were driven off in large numbers. The incursion of white men into what the Indians considered their hunting grounds was a fruitful cause of friction. But early, very early, before the hunting ground issue had been raised or considered, the Comanches and other predatory tribes of the Southwest, including Kiowas, Apaches, and Navajos, made a seasonal habit of raiding the settlements.

It follows that early Texas history is full of wars with raiding Indians and that generally the cattlemen bore the brunt of the fighting. In 1822, Stephen Austin, on the way from San Antonio to the City of Mexico to conclude his land grant negotiations, was captured and robbed by Comanches. The Father of Texas loudly proclaimed himself an American, pointing out that the Comanche quarrel was with the Mexicans. Apparently in admiration of the originality of this summation, the chief of the raiders returned Austin's property and permitted him to continue his journey unharmed. As late as 1827 the Comanches still made shift to preserve their friendship with the "Americans" as distinguished from the Mexicans in Texas. But the strain of keeping their hands off so much inviting livestock was too great. Shortly thereafter, with apparent relief, they dropped all such artificial distinctions and began to steal horses and cattle with renewed delight and fine impartiality. The respite, however, had been timely. During those early years of American settlement the Texas pioneers found themselves very much occupied with the Caranchuas and Wacos, whom they "settled" by about 1826.

A candid postscript on a letter caused the first open conflict with the Comanches. A war party of these Indians, on its way down the Brazos on the friendly errand of stealing horses from the Lipans, stopped at the ranch of Captain Sims, in Burleson County, one day in 1827, and asked for a letter introducing the bearers as "good Indians." They wished this "papel" to show the settlers farther on, and thus avoid trouble with them. A certain Tate, living at the ranch, wrote the requested letter as Captain Sims dictated, but added his own opinion: "From all appearances I am inclined to think these Indians are going down the Colorado to steal from the settlers."

The postscript was a death warrant. When the Comanches innocently showed the "papel" the settlers read and were alarmed. A hurried posse was raised by Colonel S. P. Ross, and all except two of the warriors in the band were killed in a sudden attack on their camp. The two escaped because they happened to be riding faster horses than any of the white men. On their way back to their own country this pair stopped with the praiseworthy intention of "collecting" the perfidious Tate, but their aim was unfortunately bad, and although wounded he survived and left the country in which he had stirred up a bloody and extremely cruel war. Through all the succeeding years, until nearly 1880, the Comanches were to remain constant and enthusiastic enemies of all Texans. To live on an outlying ranch became a matter of great peril. An incident which occurred in 1835 is typical.

In that year the Comanches raided the de Witt colony on the Colorado River and attacked the ranch home of a family named Hibbens. The rancher and his wife's mother were killed and scalped, and Mrs. Hibbens, with her two young children, were tied on horses and carried away. The youngest child, an infant, kept crying and the Indians presently halted and killed it. The other child, old enough to understand and obey its mother's admonitions to keep quiet, was allowed to live. A few nights later, just after the party crossed the Colorado and the raiders began to feel safe, Mrs. Hibbens, while all were sleeping, escaped from the camp. To hide her trail she waded down the stream, but the channel took a big curve, it was pitch dark, and long after she believed herself far away from her late captors she was suddenly startled and anguished by hearing, close at hand, her own child crying for her.

Here was a test for a mother's heart. It took Spartan resolution to resist the impulse to rush to her baby, but she turned resolutely away to seek help instead. Late the following day she reached a settler's home, and Captain Tumlinson, whose company of rangers was in the vicinity, took the trail of the Comanches, overtook them, killed two and captured all their horses, besides recovering, unharmed, Mrs. Hibbens' child.

So many incidents like the Hibbens raid occurred on the frontier of the Texas cow country that they became almost expected and caused little comment. Then the Texas war for independence was fought in 1835, an event so much greater than the Comanche troubles that it obscured them, although the Indians by no means ceased raiding. Out of the war for freedom came the organization of the Texas rangers—most of them cowboys—who were given the task of defending the frontier while Sam Houston's troops fought the Mexicans. Names like Big Foot Wallace, S. P. and L. S. Ross, and Ben McCulloch began to be heard in Texas.

Tough, deadly, hard, the Texas rangers were a fighting body unique in American history. At first they were chiefly transplanted Virginians, Missourians, Tennesseans and other Southerners. Theirs was the tradition of Dixie, the tradition of the fighting horseman, of chivalry, and of courage combined with amazing lethal skill. In the beginning they carried the long rifle, the horse pistol, and the knife, but after a time native Texans filled the ranks and the six-shooter became the rangers' own pet weapon. Their skill with it popularized it in the West. Organized to fight Indians, the rangers later waged war on Mexicans and wound up as hunters of criminals. As a body they were perhaps the most redoubtable man-for-man fighters on the continent.

A story is told in Borger, Texas: During the oil excitement when the town was booming there was a riot of some nature and the civic authorities telephoned anxiously for the rangers. When the next train pulled in, one solitary, lanky, sunburned, confident young man swung down to the platform.

"Where are the rangers?" asked the excited town authorities.

"I'm them," replied the young man specifically, if ungrammatically.

"Are you all that's coming?" Dismay was in the question.

"Yep."

"But we thought they'd send a company . . ."

"Why? Is there more than one riot?"

But in spite of their fighting ability the rangers were so few that it was impossible for them to patrol adequately the whole great vulnerable border which extended from the sea coast near Houston in a vast six hundred mile half-circle ending on the Rio Grande at Eagle Pass. Time after time the Comanches broke through the line, and then it was the stockmen who rallied and met them. The cattle ranches, located with an eye to the best grass and water and with a profound prejudice in favor of "elbow room," often were a day's ride apart. While this circumstance had its advantages in cattle raising, it presented definite shortcomings when Indians were on the prowl. To the latter, range cattle, even when not desired as plunder, were acceptable as easily obtained rations on the hoof, and it was fun, after a choice beef had been selected and slaughtered, to pin-cushion every other steer in the neighborhood with feathered arrows, leaving the beasts to die slowly and in agony. Meanwhile the Comanches rode off with the ranchman's horses in a cloud of dust punctuated with shrill yells.

Parker's Fort was founded by Elder John Parker on Novasota Creek, a tributary of the Brazos. It was a strong fortification, built of deeply planted posts and equipped with loopholes for defense, but it fell easily to the Comanches and Kiowas when they appeared before it May 19, 1836.

The Indians made quick butchery of nearly everyone in the fort, but the ransom racket was already showing its ugly head on the plains and the raiders carried away a few captives, including Mrs. Rachel Plummer, her fifteen-months-old son, James Pratt Plummer, Mrs. Elizabeth Kellogg, and two children of Silas M. Parker, son of Elder John Parker—Cynthia Ann, a little blond girl of nine, and her small brother John.

The two women and the little Plummer boy eventually were ransomed by friendly Indians and returned to their relatives after thoroughly unpleasant experiences. So far, however, were the Parker children taken out into the Staked Plains—that remote uplift with its high escarpment which covers most of the Texas Panhandle and extends down almost into the Big Bend country—that their whereabouts could not be discovered. Four years after the capture of the children Colonel Len Williams, accompanied by a trader named Stoal and Jack Henry, a Delaware Indian, visited the camp of the Comanche chief Paha-uka, then on the Canadian River. Colonel Williams saw a white girl in the camp and at once guessed that she was the missing Cynthia Ann, who by now was thirteen years old. He offered to ransom her, but the offer was rejected by Paha-uka who, however, granted the colonel permission to talk with her. According to the colonel's story, Cynthia Ann came to where he was standing and seated herself, Indian fashion, on the ground. He asked her many questions, but she refused to answer, perhaps from fear or because she had forgotten English, or was indifferent to his proposal to return her to her relatives.

Not long after this Cynthia Ann became the wife of Peta Nokoni, who was destined to rise to the chieftainship of the Quahada Comanches. Not for fifteen more years was she again seen by a white man. Then Victor M. Rose, with a party of white hunters, happened to visit the camp in which she lived. By this time Cynthia Ann was the mother of two halfbreed children by Peta Nokoni, and when Rose asked her if she wished to return to her people, she hung her head and said no, adding that she was happy, loved her husband and was contented with her family and children. Cynthia Ann had become a thorough Comanche.

More time passed. In 1861 Texas rangers under Captains L. S. Ross and Jack Cureton surprised Peta Nokoni's village and captured it. During the melee a man believed at first to be Peta Nokoni himself was killed by Ross, and Cynthia Ann Parker, with her youngest child, a baby girl named Topsannah, was captured. It later proved that the man killed was not Nokoni but a Mexican servant belonging to Nokoni, who had been assigned to help the women jerk some meat. The chief himself was absent on a raid with his two sons and most of the warriors of the village. One of these two sons was the later famous Quanah Parker who became a friend of President Theodore Roosevelt and was a power among the Western Oklahoma Indians in the period of adjustment after the fighting days.

Cynthia Ann Parker wept and pleaded to be permitted to return to the Comanches. Nor was she ever reconciled to being captured by the whites. After she was taken to the Texas settlements and identified by her uncle, Colonel Isaac Parker, the state legislature voted her a yearly pen-sion of $100 and appointed her uncle to be her guardian. She learned, after a fashion, to speak English and, with the habits of industry ingrained as a squaw, quickly acquired knowledge of how to spin and weave and do housework. But a quarter of a century in the freedom of a Comanche village could not be erased. Cynthia Ann, at her heart, was a true Indian.

Her descendants, the children and grandchildren of Quanah Parker now living in the Comanche country about Lawton, Oklahoma, testify to this. One of Quanah's daughters, Mrs. Neva Birdsong, of Cache, Oklahoma, a cultured woman and a graduate of Carlisle, has devoted years to studying the history of her ancestress. I spent a pleasant week end a few years ago as a guest at the rambling old house which the great chief Quanah built at the foot of the Wichita Mountains and which is now occupied by her husband and herself, and we discussed much the unanswered questions about Cynthia Ann Parker's career.

"Cynthia Ann Parker loved her husband Peta Nokoni and the boys from whom she had been separated," Mrs. Birdsong insisted. "But her pleas to be allowed to return to the wilderness were rejected. She was held a captive—an unwilling captive—with her baby. Two or three times she escaped from the white people and tried to reach the Comanches, but each time she was captured and brought back to the settlements.

"Civilized life did not agree with little Topsannah, the baby. She pined and finally died in 1864. After that Cynthia Ann, her spirit broken, a misfit among the white people, and bitter at her enforced captivity, wasted and died too. It is my belief that she starved herself to death. . . ."

Not with entire impunity did the Comanches raid, scalp and kidnap. The Texas frontier bred a potent race, a race not to be treated lightly. Many an Indian paid with his life for audacity. Again and again the rangers struck the tribes. It would be impossible even to attempt to sketch the history of that frontier conflict here. James Cox, in his *Historical and Biographical Record of the Cattle Industry*, devotes more than one hundred pages to a bare outline of it. There were so many fronts; the combats and combatants were so scattered; and few persons took the trouble to make a record of what happened—being busy with the events themselves—so that it forms a period of history disjointed, confused, and difficult to summarize.

From the time of the Mexican War to the beginning of the War between the States, the individual fights, skirmishes, ambushes, massacres and raids in Texas numbered in the hundreds. Toward the end of this period Texas cattlemen, aided by the rangers, began for the first time to make definite headway against their red enemies. One of the most important ranger campaigns started in January, 1858, as the result of a bloody Comanche raid in Erath and Comanche Counties. One hundred Texas rangers and one hundred thirty-five Tonkawa Indian scouts, led by their redoubtable chief, Placido—all under the command of Colonel John S. Ford—took up the trail of the Quahada Comanches whose head chief was Phebits Quasho (Iron Jacket), so named from a coat of mail—once worn perhaps by a Spanish man-at-arms under the flag of De Soto or Coronado—in which he was invariably attired when he went into battle. Iron Jacket was old. The real leadership of the Indians was in a younger man, Peta Nokoni, whose wife Cynthia Ann Parker was.

On the Rio Negro or False Washita, the rangers surprised and captured a small Comanche village on May 12, taking prisoner all the women and killing all the men. As the party was stealing upon the main village, however, it was seen by a Comanche scout who galloped toward his camp whooping the alarm. There was plenty of time for the Indians to prepare for the rangers whom they far outnumbered. Out from their village they streamed, armed and decked in their feathered and painted war finery, and raising the peculiar Comanche scream, which once heard can never be forgotten. In a vividly-hued line of battle they formed, shaking their weapons and challenging the small white force to come on. Iron Jacket, in his coat of mail, with shield, bow and lance, topped off with a beautiful cascading eagle feather headdress, galloped his war pony between the opposing battle lines, delivering taunts and challenges to his enemies.

It was an old trick of Iron Jacket's and one which had done much to win him his chieftainship, for his people considered him invulnerable. Several rangers shot at him, but in spite of their ordinarily superb marksmanship none of them wounded him. It began to appear that he was indeed invulnerable; but then he carried his good fortune too far. The irony of it was that not even a ranger, but a miserable scout of the Tonkawas—despised by the Comanches—bearing the unromantic name of Jim Pockmark, who caught his sights against the body of the mailed chief. The bullet crashed through the ancient steel links and Iron Jacket toppled from his saddle.

Sensing the psychology of the moment, the rangers attacked furiously. The Comanches, shaken by the fall of their leader, wavered. Right through their own village the Indians were driven, although they had four hundred warriors to Ford's two hundred thirty-five fighters.

But aid was coming for the tribesmen. From a still more distant camp Peta Nokoni, with five hundred fresh warriors, suddenly appeared. For a time the battle with this new enemy was fiercer than before, but the rangers took to cover and brought their immense superiority in riflery into play. Then a party stole through the undergrowth at one side and outflanked the Indians. The Comanches retreated; only the skillful leadership of Peta Nokoni prevented a complete and disastrous rout. At best he saved with difficulty his own camp with the women and children. In this battle the Comanches lost seventy-five dead, including Iron Jacket, whose coat of mail fell into the hands of the hated

Tonkawas. The ranger loss was small—two killed and five or six wounded.

Annexation of Texas by the United States brought a new force into the struggle with the Indians—the army. Again and again in the years before the War between the States, the soldiers took the field against the Comanches and inflicted defeats on them. One of the most notable was by Major Earl Van Dorn, who marched against the Comanches in April, 1859, and in May struck them in what is now Ford County, Kansas. The Comanches were badly smashed up, losing forty-nine killed and thirty-six prisoners. This is not surprising when the array against them is considered. Among the officers were Major Van Dorn, Lieutenant William B. Royall, Lieutenant Fitzhugh Lee, Lieutenant Manning M. Kimmel, and Lieutenant George B. Crosby. There was a young trumpeter in the detachment named Edwin M. Hayes. Every one of these men became general officers in either the Confederate or federal forces during the war which soon was to break out between the states. Fitzhugh Lee was severely wounded, an arrow being driven through his body. The Comanches were to be pardoned for retreating before such a combination of military genius as faced them that day.

Even a struggle as fierce as that with the Comanches on the Texas frontier had its lighter moments. A cowboy on the Clear Fork of the Brazos, whose name unfortunately has been lost, was the hero of one such incident. Line riding one day, he found a cow with an unbranded calf, roped the little creature as a good cowboy should do, built a fire, heated an iron, and placed his employer's brand on the calf. At this juncture a solitary Indian lurking in the brush near at hand took a pot-shot at him.

Without an instant's hesitation the cowboy vaulted into his saddle, swung his rope and went at the savage with a yell. Vainly the Comanche tried to load his old-fashioned gun, saw vengeance bearing down upon him, and turned to run just as the noose whistled through the air, whipped about the Indian's legs, and brought him to the ground. The well-trained cow horse immediately began to walk sedately toward the branding fire which was still blazing, dragging behind, as if he had been any ordinary maverick, a very glum and sullen captive.

At the fire the cowboy dismounted, reheated his iron, and grimly placed his outfit's brand on a prominent portion of the Comanche's anatomy which was left bare by the dangling breech clout. Then, snatching the Indian's own knife from its sheath, he bent over the prostrate and bedraggled savage and added the ranch earmarks as corroborative proof of the authenticity of the brand. Having completed all this he loosed his rope and kicked his captive to his feet.

"Git now," growled the cowboy, "an' the next time some feller calls ye a maverick, tell him that he's a liar. An' tell yore folks, when ye next see 'em, that I'm coming no'th, come next grass, to round up an' brand the hull damned tribe."

V

HIDE AND TALLOW EMPIRE

While rangers, Indians and stockmen were circling and sparring for holds on the frontier, farther back in the semi-settled areas of Texas the cattle country rapidly expanded. Most American settlers who went to Texas became stockmen—and for a good reason. The Spanish government, to encourage colonization, had established a policy of making land grants, and this policy the Mexican government continued in Texas up to the revolt of that volcanic young state.

The peculiarity of the Mexican grants was this: American settlers quickly learned that if a man declared his intention to farm, he would receive one *labor* (177 acres) of land. If, on the other hand, the same man declared his intention to raise livestock, he received a league (4,438 acres) or more. This was not due to any desire of the Mexican government to favor stockmen over husbandmen, but was based rather on a consideration of what land a farmer could till to crops as contrasted to that which a rancher could utilize. Naturally, however, any man with good sense—and those first Texas settlers did not lack in canniness—made the declaration which obtained for him the larger acreage. All the more natural was this since Texas was largely settled by Southerners, to whom the free, active life of cattle ranching, with its horsemanship, action and drama, appealed far more than sober grubbing on small acres. There were, to be sure, several large areas taken up by cotton farmers, but these were of a different *genre*.

The logical outcome of this policy was to shape the entire economic life of Texas by the time it entered the United States, as is shown by tax figures. Taxes were paid on the following numbers of cattle, not including "wild cattle," in the years indicated:

1848	382,873
1855	1,363,688
1860	3,786,443

Texas had more cattle than she knew what to do with. In 1833 there were approximately twenty-one thousand persons living in the civilized parts of Texas—that portion covering only a strip along the southern border, from Natchitoches, Louisiana, westward to where Eagle Pass now stands on the Rio Grande. At the same date there were about one hundred thousand cattle in Texas—fifty thousand in the Nacogdoches department, and twenty-five thousand each in the Brazos and Bexar departments. That figures out to five head of cattle for every resident of the Texas settlements, and later this ratio rose to six cattle for every human being. In the rest of the United States at this time there was less than one head of cattle per person.

Texas, however, was cattle poor. So cheap were long-horns that not a few cattlemen thought it hardly worth while to round up and brand their new calves each year. Often a mature steer commanded a price of less than four dollars. By 1845 the cow country was almost prostrate. It was at this period that the peculiar and distinctive word "maverick" became an integral part, which it still remains, of the cattle language. Endless variations of how this came about exist in the West. McCoy, in *Historic Sketches of the Cattle Trade*, written in 1874, offers the following account:

Unbranded animals over a year old are, in ranchman's par-lance, called "mauvrics," which name they got from a certain old Frenchman of that name who began stock raising with a very few head, and in a very brief space of time had a re-markably large herd of cattle. It was found that he actually branded fifty annually for each cow he owned. Of course he captured unbranded yearlings.

Another version is that Colonel Samuel A. Maverick of San Antonio went off to the War between the States, leaving his cattle to take care of themselves. When he returned from the Confederate army, he discovered his fellow ex-soldiers had helped themselves liberally to the unbranded cattle on the Maverick range, so much so that an unbranded adult animal thenceforth bore the name maverick.

A third account is that the stockmen of Colonel Maverick's district before the War decided to form a cattlemen's pool with all brands to be registered. The legendary colonel, according to this story, spoke to his fellow ranchers in words somewhat as follows:

"Wall, now, since all yo' gentlemen have chose yo' brands, I'll jes' let my brand be nothin' at all. Hyarafter yo'-all will know that a cow with no brand is a Maverick."

The ridiculous part of this story is that the other cattlemen are supposed to have been taken in by the transparent artifice and permitted Maverick for a season or so to fall heir, without turning a hand, to every calf which strayed from its mother or was overlooked by the branding crews.

Probably the best authenticated story is the following:

In 1845 Colonel Samuel A. Maverick of San Antonio, a lawyer rather than a cattleman, received in payment of a legal fee four hundred longhorn cattle. The stock could not be sold since there was no market for them, so the colonel sent the herd down to his ranch, forty miles south of San Antonio, where he put them in charge of his slaves. The colonel's blacks, however, were not celebrated for their energy. The trouble and labor of branding did not appeal to them, and since the colonel was not present himself to supervise, very few of the herd increase ever saw the iron, most of them running wild and unmarked until they matured.

After a time Colonel Maverick, to his surprise and gratification, found a buyer for his cattle, Toutant Beauregard. When the latter went to take possession he found that Maverick's cattle were unmarked and immediately made

claim for and branded to himself all unmarked cattle on
the surrounding range, from which incident an unbranded
beef animal, old enough to leave its mother, is today known
as a maverick from Kingsville to Calgary. As unclaimed
property running loose on the range, requiring only a hot
brand to establish title, mavericks were and are constant
sources of temptation and generators of friction. Struthers
Burt has said, "The word 'maverick' killed more men in
the Far West than any other word."

As to the superabundance of Texas cattle and their con-
sequent almost disappearing value, it is a matter for wonder
how the owners of the longhorn ranches survived at this
period. Stephen Austin, in his statement of affairs in Texas
to the Mexican government about 1833, wrote:

The raising of cattle (Texas longhorns) and hogs (razor-
backs) has increased with so much rapidity that it is difficult
to form a calculation of their number. The price for which
they sell will give you an idea of their abundance. Fat beeves
of from twenty to thirty *arrobas* are worth from $8 to $10.

When it is understood that an *arroba* is about twenty-
five pounds, it will be seen that a 750-pound steer in good
flesh was selling for ten dollars. Cattle raising in Texas,
however, was not an expensive process. Grass was free and
plentiful. The climate was so mild that the stock required
little attention. Without stimulation the cattle increased.
Owning beef in abundance but with no means of obtaining
an income from it, many ranchers worked to create their
living out of the soil. Ranches had their "garden patches"
where potatoes, corn, peas, beans and cabbages were
grown. A few sheep were generally kept for wool, and
every woman was expert in carding, spinning and weaving.
"Homespun" was familiar to every Texan as clothing material.

A ray of light appeared now in the economic gloom. A
market began to grow up along the gulf coast for hides,

tallow, hoofs and horns. Not yet had the world discovered the use of electricity for lighting, and even natural gas had not been utilized. The coal oil lamp was an expensive luxury. In most homes "tallow dips" were the common form of illumination. Texas sold great quantities of tallow, rendered out of the carcasses of its longhorn steers, to the rest of the world for candlemaking. In the East manufacturers made candles by machinery and sold them boxed in lots of forty, but many a housewife thriftily dipped or molded her own candles. Hides also began to sell, cheaply yet steadily, to buyers from the Atlantic seaboard looking for leather material to make harness, shoes, saddles and other articles. Hoofs of slaughtered cattle found buyers among glue manufacturers, who also purchased horns; and a quantity of hair, even, was disposed of for plaster and padding. In fact, it seemed there was a market for everything about a steer except his beef.

Small packeries known as "hide and tallow factories" sprang up. Some of them "salt junked" meat and sold it in Mexico or New Orleans, but most frequently after hide and tallow were removed the carcasses were left in heaps to rot or be devoured by buzzards, coyotes and other carrion eaters. One of the original packeries was operated before the War by Captain Richard King, who with Captain Mifflin Kenedy established the great King ranch in the southern tip of Texas, which today is still one of the largest in the world. King tried to preserve beef on a theory of his own—by infusing brine into the veins of the carcasses. The experiment in crude embalming failed and thereafter the captain devoted his establishment to preserving the hide and tallow only.

Such beef as was preserved by the old packeries usually was pickled in barrels. That there was some success with it is shown by the fact that in 1844 it was quoted on the market at Galveston as a staple item at $10 a barrel. One venture in packing deserves special mention. In about 1851 the foundation of a great fortune was laid in Texas when

Gail Borden, who became many times a millionaire by condensing milk in tin cans, obtained his first stake. He did it by putting up "meat biscuit"—tinned beef—for the trade at Galveston.

Texas was really preceded by California in the hide-and-tallow trade, and throughout the Southwest a steer hide was known as a "California banknote." Many coasting vessels, including the ship which carried Richard Henry Dana when he wrote *Two Years Before the Mast*, visited California for hides. The trade there, particularly in the earlier days, was largely on a barter basis, ship masters carrying large quantities of cutlery, cloth, and other articles to trade for hides and tallow, while many ships had their own equipment, including huge brass kettles, for rendering out the tallow.

The hide and tallow factories furnished a market, such as it was, but it was woefully inadequate for Texas, with its immense cattle surplus. Inevitably the cattlemen sought an outlet somewhere else. As early as 1842 a rancher drove a herd of beeves to New Orleans. A few herds found their way to Shreveport whence they were taken down the Red River in steamboats. One of the Shreveport trail drivers was John Chisum, of whom more later. It was north, however, that the market logically had to go. That cattle owners should begin feeling their way in that direction was a question of time only. The first trail herd which is authenticated as making the march toward the Pole star from Texas was driven in 1846 by Edward Piper up through Missouri to Ohio, where he arrived with most of his herd, which he fed up fat and sold at a good round profit.

And there is a strange story, shrouded now in complete absence of details, which eclipses every other trail drive episode of this period. It is contained in a newspaper item quoted by the United States Department of Agriculture in 1855, which mentions that a drove of several hundred cattle from Texas was passing through Indiana County,

Pennsylvania, on the way to New York City. One wonders who the adventurers were who made that epic journey. Surely theirs must have been a fabulous experience, if we could learn the full story. How long were they on the trail? What was the purpose of that long Odyssey when other markets were nearer? What became of the animals? History is silent. Only that bald statement from an ancient newspaper brings them for an instant vividly to life, marching steadfastly with their long, horn-clashing procession, toward the great blue sea, their simple Texas minds no doubt amazed by all that they saw, but their rugged Texas courage unruffled by any perils through which they passed.

From 1845, the year of the great cattle depression, until the opening of the War between the States, trail driving increased. A tremendous stimulant was the California gold rush of 1849. That unexampled stampede of a whole population to the Pacific coast caused California to face for the first time in her history a shortage of livestock. A few daring Texans began driving cattle to San Francisco, feeling their way across the desert, fighting off Apaches, hunting water. Always hunting water. One of those California trail drivers was W. H. Snyder. It took his outfit more than two years to cover the route, which followed the Rio Grande north to the mountains, crossed the continental divide in southern Colorado, crossed a corner of Wyoming and northern Utah, and took the well-marked California overland route from there, angling southwest across the Nevada alkali flats into California and thence to San Francisco. It was a prodigious, almost an incredible feat . . . but here again we are faced by the inarticulate cowboy's failure to leave a record. Snyder and his trail drivers were too busy with their work to take time to write about it.

A few details were given by W. A. Peril, an old-timer who drove the California route, over part at least of Snyder's trail. On one drive Peril's outfit wintered at Brown's

Hole on the Green River, and two bands of Ute Indians camped close to the cowboy's camp throughout the winter. When the snow melted in the passes and the trail opened the following spring, the cattle were set in motion, passing Fort Bridger, thence down the Bear River through Bear Lake Valley, Soda Springs, and on down the Snake River to where the Oregon and California trails divided. At that point the herd took the south fork, crossed Portneff, Goose Creek and Raft River, passed through the City of Rocks and Thousand Springs Valley to Humboldt Wells and then on down the Humboldt River to Lassen's Meadows, where the cattle were pastured until they could be sold to the mining camps.

Another way to California was the so-called "Southern Route" . . . but men with a lingering prejudice in favor of preserving the hair on their heads usually avoided it. Athwart that trail lay the Apaches and the scalp sat loose on the skull of the man with the temerity to travel it. Even so a few Texas drovers had the cold nerve to make the attempt. One of them was Captain Jack Cureton of the Texas rangers, accompanied by his son, W. E. Cureton. They formed an outfit which took eleven hundred longhorns over the Rio Grande at Albuquerque, followed the Little Colorado River across part of Arizona, then cut across to California where they trailed the herd up the barren eastern escarpment of the mountains to a pass across the Sierra Nevadas into California's lush pastures. Captain Cureton paid ten dollars a head for his longhorns in Texas, and sold them for thirty dollars a head in California. That gave him a neat profit even after deducting strays and deaths among his cattle, but most men—even Texans—were willing to yield to him the exclusive right to the route as long as Mangas Coloradas, Cuchillo Negro, Delgadito, and the rest of the Apaches cared to dispute its use. Some few who did not subscribe to this feeling salted the desert with their bones.

Such were the uncertain predecessors of the great migrations of the cattle herds. The first drives were sporadic, seeking blindly, this way and that, for a market and an outlet. New Orleans and Mexico were tried by some cattlemen. California, Missouri, Ohio and even New York by others. It was the beginning of a great impetuous movement. Before the spate of trail driving could arrive, however, a war intervened its bloody interlude.

VI

THE COWBOY YELL GOES TO WAR

His full name was Finis V. Ewing, a name held in esteem
by his neighbors, and which, when signed to a check, was
accepted without question at its face value in any bank
from San Antonio to Kansas City that knew anything
about the cattle business. But everybody called him "Fine,"
and fine he was . . . a stocky, graying man, with a crisp
mustache, a great fund of humor and, to the end, the out-
spoken convictions of an unreconstructed Confederate.

During the '80's Fine Ewing owned a cow ranch on
Mule Creek, about twenty miles south of Kiowa, Kansas,
in the Indian Territory, and he went on occasion to town
to transact business. When he sold his cattle, it was his custom to celebrate the event with sufficient hard liquor to
"cut the dust in his throat." Always Fine Ewing took his
whiskey standing, that being his method of determining at
what point he had consumed enough. As the evening
lengthened, he would be impelled to recount his life's history, always in the same concise words:

"I was bawn in Missoury, raised in Californy, eddicated
in Texas, gradyouated from the Confed'rit army, an' now
I'm practicin' in the Cherokee Strip!"

This summary of his career usually occurred to him at about the time he noticed that he was weaving on his feet. At this point he would send for his horse, not scorning assistance into his saddle. Once mounted he became steady as a rock. No trace of instability could be detected in him, or of inebriation either, except for the fact that his shrewd gray eye was a trifle brighter and more jovial than usual. Having seated himself in his tree, he was wont at such junctures to drive his spurs into his horse, wave his wide hat exultantly over his gray head, and go surging out of town in a cloud of dust, with a shrill wolf cry cutting the air behind him:

"*Yip-Yip-Yip-e-e-e-e-e-e-e-e! Yah-ah-ah-Yah-e-e-e-e-e-e-e!*"

Fine Ewing, who had ridden with Jo Shelby's raiding Missouri cavalry in the Confederate army for four years, said that it was the Rebel yell, and a good few who heard it endorsed the identification. But it was also—and to this Fine Ewing agreed—a cowboy yell. And that was the thing about it which was of some significance.

Today when the band strikes up "Dixie" in the South, you can hear, usually, the Rebel yell, and when it is properly given there is about it something familiar to anyone who ever has been in the cow country. The Rebel yell is of the warp and woof of America's history. At Bull Run, Gettysburg, Shiloh, Atlanta, Chancellorsville, and hundreds of other battles and skirmishes in the long-drawn War between the States, its shrill, tympanum-piercing howl of exultation was as much a part of those conflicts as the thunder of the guns.

It is my belief that the Rebel yell originally was a cowboy yell; that it came from Texas with the thousands of young men who quit their cattle ranches to join the Southern armies, punctuating the air with shrill yips, keyed to the wild challenge of the coyotes; and that this yell, picked up and quickly adopted by the whole of the gallant host in butternut gray, became stamped as the peculiar defiance of the Lost Cause.

Texas, in the War between the States, was devotedly, stanchly for the South. The flower of the state joined the Confederate army, and the record of the Texas regiments was notable in a war filled with notable feats of arms. Some of them fought under Lee, others marched with Johnston, Bragg and Hood; and still others engaged in the far-flung and little-known theater of war on the border, where Shelby and Marmaduke and Kirby Smith operated. There, had there been an adequate chronicler, an epic might have written itself into history.

The young men of Texas eagerly rode off to the colors and the cattle herds were left in charge of women, old men, and young boys. And at once Texas bestirred herself to provide not only fighting men, but food for the gray armies. In the cow country, trail drivers, many of them youths so young that the down scarcely sprouted on their cheeks, turned the noses of wild, wide-horned cattle toward the Mississippi River. They did not know the country to which they were going, but they knew that over there somewhere the troops of "Marse Robert" and other valiant leaders of the gray were in need of beef. The risks they were willing to undertake; such profits as might be made were acceptable; and the adventure was an end in itself.

W. D. H. Saunders was a lanky boy of seventeen in October, 1862. He had the slim recklessness and the lurking light of deviltry which youth carries in its eyes, and he was riding out from Goliad, Texas, on a great hegira. With Jim Borroum and Monroe Choate and a few cowboys he was starting a drive with eight hundred longhorn steers, the objective being New Orleans where the herd would be turned over to the Confederate quartermasters. News traveled slowly in those days, particularly among the widely scattered ranches of the cow country. Therefore the boy trail drivers, starting out so confidently from Goliad, did not know as they strung out their steers for the east that New Orleans was already in the hands of the Yankees and had been since April.

At Clinton they crossed the Guadalupe River without mishap and thus were launched on what is, to my notion, one of three or four of the most remarkable drives in range history. In Lavaca County their cattle stampeded, but stampedes were to be expected with animals as wild as these, and with philosophy and patience the Texans spent several days rounding up the scattered herd. Undaunted by the occurrence, once more they resumed the drive, swam the steers across the Colorado at Columbus, the Brazos at near Richmond, the Trinity at Liberty, the Natches at Beaumont, the Sabine near Orange, and reached Louisiana where they passed through Opelousas, a noted cattle market in the early days.

There they encountered two other trail drivers named Crump and Fleming, with a small herd, which was pooled with that of Saunders and his partners for the remainder of the drive. The new cattle swelled the total of the drove to about eleven hundred. On went the combined outfits, keeping out a wary eye for Yankees. But it was not Yankees they needed to worry about. A detachment of cavalry rode up one morning—in gray uniforms. The officer in command regarded the Texans and their herd sternly.

"Howdy," said Saunders.

"Whar yo'-all going?" asked the officer.

"We're headin' for New Orleans."

"New Orleans, eh?" said the officer grimly. He snapped an order and the trail drivers found themselves arrested by the detachment. Moreover, they heard themselves being charged with aiding and abetting the Yankees by transporting beef to them!

Indignation and amazement struggled in the faces of the Texas youths. It was their first intimation that New Orleans was not held by the Confederacy. In spite of their protests that they would rather have been shot than help the hated Northerners, a drum-head courtmartial was ordered at Alexandria, Louisiana. The presiding officer at the court must have been a man of judgment and a reader of human character, for he believed the story of the boys and released them to continue their drive.

"Stay a good piece to the no'th of N'Awleans," was his parting injunction.

The New Orleans market was cut off. Where should Saunders and his partners take their cattle now? Their inquiries satisfied them that Mobile was still in the possession of the South, and although it was much farther on, and on the other side of the Mississippi, the young Texans thought they might sell the steers there. A few days later they were on the banks of the Father of Waters, and a terrific problem presented itself. There were no steamboats to ferry the cattle across. How were they to reach the other side with their herd?

"Never know if you kin do a thing onless you try it," was the dare-devil philosophy of the trail drivers. Right down to the brink of the rolling Mississippi, a mile wide and forty feet deep at that point, the wild young cowmen took the steers—and then with shrill yells and lashing lariats urged the animals into the water. One boy took to the river on his swimming horse and rode ahead of the herd as a guide. Others flanked the cattle which by now were churning the river into a creamy foam. The rest brought up the drag. It was a manifest impossibility to swim a herd across the lower Mississippi—yet somehow these Texans did it. Of the eleven hundred Texas steers, a full thousand took to the water like amphibians and swam the mile-wide torrent, landing far downstream on the other side. The herders—mere children, some of them, their milk-teeth scarcely shed and their beards ungrown—swam their horses alongside and whistled and howled at the cattle like exultant young fiends. It was the Rebel yell they were giving, although they did not know it, and the longhorns, their eyes rolling in fear and panic, wallowed in the waves like hippopotami, breasted the muddy tide, struck out boldly, each following the wake of the beast in front of him, and eventually, miraculously, clambered in safety up the opposite bank of America's greatest river.

About a hundred of the cattle refused to attempt the current and these malingerers were sold on the west bank to somebody, possibly a Confederate quartermaster. The thousand which swam across were driven all the way through Louisiana toward Woodville, Mississippi. And when the state line was crossed, more trouble lay ahead of the cattle argonauts.

Only a few days after they performed the prodigy of swimming their herd across the Mississippi, while they were trailing along in peace of mind and high spirits, again a dusty file of gray-clad horsemen clattered up to them. Another officious Confederate provost marshal placed them under arrest on suspicion of attempting to aid the enemy. But this time Saunders and his companions did not win release as quickly as before. They were taken to Fort Hudson and there forced to spend several days in a guardhouse before once again they were able to convince a courtmartial of their loyalty to the cause. After their hearing, however, they were quickly released and once more allowed to resume the much interrupted drive—their cattle, apparently, having been obligingly herded for them by Rebel soldiers in the meantime. So at last they reached Woodville. There Crump and Fleming cut out their cattle and took them on to Mobile while Saunders and his friends sold their share at Woodville.

And did the somewhat inhospitable treatment accorded them cool the Confederate enthusiasm of the Texans? Not to a noticeable degree. As soon as his cattle were sold, Saunders returned to Goliad, hurriedly settled his affairs and went to Corpus Christi, where, on February 23, 1863, he enlisted in the Rebel army.

Other Texans likewise made the long journey east to drive cattle to the Confederacy, although it is not recorded that any of them duplicated Saunders' feat of swimming a herd across the Mississippi at its widest and deepest point. Branch Isbell, later himself a Texas trail driver, was a boy in Alabama during the War. He used to tell of seeing Confederate soldiers driving herds of Texas steers as far east as that, and the song they sang stuck in his memory:

> Driving cattle's our promotion,
>   Which just exactly suits my notion,
> And we perform with great devotion,
>   There's work enough for all.
>
> I'd like to be a Virginia picket,
>   But I'd rather be in the cattle thicket
> Where the hooting owl and the screaming cricket
>   Make noise enough for all.

Of the Texas drivers of cattle to supply the Confederacy, probably the most tireless was Colonel D. H. Snyder, who was born in Mississippi in 1833, "the year that the stars fell." In the second year of the War Snyder took charge of a beef herd contracted by Terrel Jackson, a wealthy Texas planter, to be delivered to the commissary department of the Confederate army. Snyder, an experienced cowman and a born leader of men, was herd boss. Through difficult and dangerous country he took his steers to the Mississippi somewhere near Vicksburg, where the herd was delivered to Major Ward of the Confederate quartermaster's department and presumably ferried across the big river. Snyder returned at once to Texas and delivered several more herds to the Southern armies.

An innovation of Snyder's was his trained swimming steers. Always it was difficult and dangerous to cross a river with a herd, but the problem was much simplified if the cattle had leaders to show the way across. Ordinarily a cowboy swam his horse ahead of the herd to set the example, but Snyder improved on this method. He trained two steers as swimmers. So well taught were the beasts that when they came to a river they would plunge boldly into it and strike out for the opposite shore. Other cattle blindly followed the lead, and so the herd would cross. In the cattle country Snyder's "lead steers" became celebrated, and they served him many times before he was forced to quit trail driving to the Mississippi because of the fall of Vicksburg and the Yankee gunboats prowling up and down the stream.

To have swimming steers was an idea so good it is a wonder other trail drivers in later years failed to copy it. Perhaps Snyder never let it be known how he taught his "lead steers" to lead the swimming. Charles Goodnight in later years had a trail leader named Old Blue, but he was not a swimming steer. Snyder's unique expedient of the trail passed into history when the trail driver, after Grant's victories closed the river, slipped across it one night in a skiff, and, unable longer to bring beef to the Confederacy, offered himself as a soldier in its ranks. He was as brilliant a soldier as he was a brilliant trail driver and at the end of the War possessed a colonel's commission in the hard-fighting gray army.

When the short, square-bearded Yankee leader, Ulysses S. Grant, captured Vicksburg in July, 1863, he completed the severing of the Confederacy along the line of the Mississippi, and ended cattle driving from Texas. Thereafter the gray forces slowly starved until they surrendered at Appomattox.

So the great war dragged on east of the Mississippi, and fecund Texas, to the west, with no market for her cattle, saw her herds grow in almost geometrical progression. In the late years of the conflict the saying was that "a man's poverty is measured by the number of cattle he possesses." After peace came, ranchmen who returned home found their herds roaming wild and so valueless that the rounding up and branding of them were almost a waste of time and money. Cattle were known to have been sold for as little as two dollars a head. Land was almost worthless and the South was in the throes of the reconstruction. Texas, which had suffered less from the actual ravages of war than other states in the Confederacy, was financially as prostrate as any of them. Even those venturesome trail drivers who had succeeded before the blockade of the Mississippi in disposing of cattle across the river had nothing to show for it but Confederate currency—as worthless now as dead leaves.

Yet the dark cloud developed a silver lining. Out of the great forces loosed by the War between the States grew the very market which Texas had needed all the preceding years. In several ways events growing out of the struggle conspired to help the cattle country. After an exhaustive study of census records for the United States, covering the years 1840, 1850, and 1860, Professor Silas Loomis estimated in 1865 that the country required eighty cattle to supply the needs of each one hundred people, and that this ratio must remain constant if the needs of the population were to be met. Pursuing his thesis, he found that in 1860 there was a deficiency of 785,161 cattle in the New England states, 1,564,526 in the Middle states, and that the South, excluding Texas, was about normal. In 1860, as Professor Loomis reported to Congress, Texas owned 438 cattle to every one hundred persons, although Massachusetts had only one-fourth enough to support her population, and New York, Pennsylvania, New Jersey, Delaware, Maryland, and Virginia, all lacked enough cattle to care for the needs of their citizens.

These were the conditions before the War. They became vastly more disjointed as the War raged on its destructive course. In the North, mammoth army orders caused government contractors to comb every available section for beef. At the same time production of cattle fell off, owing to the fact that so many men, a number of them stockmen, were in the ranks. A great change in emphasis, moreover, was occurring in agriculture. In the decade prior to the opening of the War, Cyrus McCormick had brought out his wheat reaper, and other labor-saving machinery for agriculture had been developed. The farmers left at home discovered that by the aid of the new machinery they could produce wheat almost as well and as fast as when the full population of farm workers was available. Wheat was a quick money crop. One season would produce a profit on wheat, while it took three or four years to grow cattle to a marketable stage. What with the new machinery, moreover, there was not the need for the labor and constant care in wheat farming which was required in stock raising.

It was only natural that more wheat should be grown in the Eastern and Middle states. And in the settlements in Kansas, Iowa and Minnesota, practically all farming efforts were devoted to wheat and corn. A great natural cattle range was thus lost, since under the conditions of that day cattle raising was essentially a frontier industry and the frontier was turning its attention elsewhere. This combination of factors produced a disastrous beef shortage in the North by the end of the War. Prices rose to prohibitive heights not only for beef but for other foodstuffs, and Northern artisans were unable to afford meat for their families.

Other results of the War combined with the beef shortage to set the stage for the tremendous spectacle of the cattle drives. An industrial revolution had come about. War orders, greatly stimulating manufacturing, brought tens of thousands of people, who would otherwise have been producing food on farms, to industrial cities. Another factor was as important as any of these. The nation's transportation consciousness had been greatly stimulated by the War. In the military campaigns the railroad played a part so vital that its importance was dramatically emphasized. It was becoming apparent, moreover, that the Pacific coast and the East must be linked. As early as 1862 Congress passed the Union Pacific act, offering national assistance to the proposed railroad which was to be built across the continent; but the War prevented construction of more than a few miles of this line.

To stimulate further railroad building, Congress now hastened to pass other laws. Ten sections of public lands and the loan of sixteen thousand dollars of United States bonds per mile of construction were included in the earlier provisions, and these terms were made even more inviting by later revisions. Stimulated by these inducements and by a hot spirit of rivalry, the construction race across the plains began by 1866. Gangs of Irish immigrants pushed the steel westward from the Missouri, and gangs of Chinese coolies pushed it eastward from the Pacific.

Transportation was to be offered to the West by the Iron Horse, and Texas was destined to be the first to profit by that transportation.

2. . . . HELL AND HIGH WATER

VII

CHICAGO GREASE AND THE WESTERN EPIC

Back of Chicago's stockyards, shortly after the War between the States, ran a filthy, malodorous open sewer known as Bubbly Creek. Men shunned the place as a general rule, yet every day or so an elderly gentleman visited Bubbly Creek with a hawk-nosed curiosity. He was a tall, lean man with a gray beard, and he wore a decent, well-brushed black suit with tail coat and stiff hat, the trousers stuffed into coarse cowhide boots.

The bearded man was G. F. Swift, one of the pioneers of the meat-packing industry, and his reason for visiting the noisome sump was a basic element of the new movement which was changing the packing industry and was destined to remake the face of the West. He went to stinking Bubbly Creek, relates his son, Louis F. Swift, in his biography *A Yankee of the Yards*, to find where his company's sewer emptied. Over that outlet he bent and studied it with deep concentration. He was peering down to detect any traces of fat which might be coming forth. If he saw any, the long legs took him rapidly to his superintendent's office, and a rigid check-up followed. Fat, to Swift, meant waste; and elimination of waste was the beginning, the middle, and the end of the new credo of the meat-packing industry, at that period growing mightily in Chicago and later spreading to Kansas City and other Western centers.

The Union Stockyards at Chicago were completed and opened for use on Christmas Day, 1865. From that hour the cattle empire may date its real growth. P. D. Armour and John Plankinton were two Milwaukee pork packers who had been far-sighted and lucky enough to bet on Grant and the Union army by selling pork short during the War between the States. They cleaned up a million and a half dollars and went down to Chicago to establish a beef-packing business close to the Western source of supply. Armour was a square-set, side-whiskered man, one of the great organizing geniuses of a generation of remarkable organizers, whose secret pride was not his business achievements, but his reputation as a creator of aphorisms. Some of his sayings have come down to us. *Experience is cheap at any price if you benefit by it* was one. Also: *Brains always have and always will command the highest market value.* He summarized his own character thus: *Most men talk too much. Most of my success has been due to keeping my mouth shut.*

These sayings do not sound very brilliant today, but Mr. Armour employed many people and doubtless did not lack for admiration. Nor did he lack tremendous native force and ability, and that was why the Armour interests took so important a place in the packing world. Armour and Plankinton led the way for the packers in the West. Swift and others followed them. The rise of the packing business and the growth of trail driving were simultaneous and interdependent.

Gustavus Franklin Swift was a Cape Cod Yankee and his story is interesting because it reveals much of what took place in those early days of the modernization of meat packing. There has been some dispute as to who coined the phrase about using everything of the hog but his squeal, but Swift has as good a claim to it as anybody and it was the keynote of his character and policy. His preoccupation with the elimination of waste came from his study of butchering back in Massachusetts where he had started with almost nothing and built up a prosperous meat business before going west to Chicago. When a beef was

slaughtered it was the New England custom to save, besides the meat and hide, only the head, feet, tripe, heart,
liver and tongue. Head, feet, and tripe were called a "set"
and sold separately. A customer who purchased a carcass
entire received gratis a heart, liver and tongue.

It was an old custom, ordained by generations of tradi-

tion. But Swift, who disliked intensely giving away anything, observed that near the big slaughterhouses were
small establishments where people, taking advantage of the
loose ways of the slaughterers, made a living on what the
butchers threw away. They were jackals of the trade, who
came humbly to volunteer to haul away the offal, and the
big butchers, happy to be rid of the "rubbish," easily gave
consent. But Swift observed that the offal was taken to the
processing places of the scavengers and there was refined
into such products as neat's foot oil, pickled tripe, soap,
and other by-products, which enabled the garbage collec-

tors to make a profitable business out of material for which they paid not one cent because it had no market value. Swift studied deeply the problem of changing that.

Out in Chicago, meantime, the Union Stockyards had suddenly burgeoned into a great cattle market. Furnishing meat to the population of a city was originally a local task, and every town of any size possessed a slaughterhouse of its own. In cities like St. Louis, Cincinnati, and Louisville, these sometimes were quite extensive. Cincinnati, for a period, was a meat-shipping center of importance, dealing in pork with the South, which furnished a market because its large rural population was devoted to cotton growing and had comparatively little livestock. For a long time before Chicago superseded it, Cincinnati was known as "Porkopolis." One of the objections to the slaughterhouse system was that as cities grew their business or residential districts continually encroached upon the stockyards, compelling them to move. In Chicago, at last, a group of speculators purchased half a section of land on Halstead Street, combined all the dealers' interests, and built the Union Stockyards. This occurred at an ideal time. The next spring saw the first of the big cattle drives from Texas.

To Swift these events were inspiring. Born and reared on the narrow hook of Cape Cod, which juts out from the shore of Massachusetts, and from which hail the most typical Yankees in all New England, he was at fourteen working in his brother's butcher shop, and a few years later had a shop of his own, at which time his thrifty Yankee nature began pondering the wastes described above.

"If other men can make a living off my waste," he reasoned, "I must be losing money in some way."

As Swift progressed through the various stages of butcher, local dealer, export cattle shipper, and wholesale meat dealer, his theories began to take form. A great inspiration arrived when the first important rush of western cattle began. At that time his business was already widespread in the East—at Albany, at Buffalo, and in Massachusetts. To sell beef there he had to ship in cattle, and the best place to buy cattle was in the West. Always figuring and comparing, the Cape Codder discovered that when he bought western cattle which had passed through the hands of two or three middlemen, he had to pay extra prices for them. Other men also had arrived at this conclusion, but Swift acted on it. He went west himself to buy and ship direct from Chicago to his eastern plants.

Now came another step in an interesting progression of ideas. Only about forty percent of a steer was edible, Swift computed. Clearly, therefore, it was sinful waste to pay freight on the remaining inedible sixty percent of the animal. Moreover, he became convinced that he lost value on his live steers because they frequently became bruised on the long train ride east, and almost invariably lost weight —"shrank" was the cattle buyer's phrase. There was, besides, the expense of feeding and caring for the cattle on the way. It was an adding up, a mounting, of wastes one on top of another.

The logic of Swift soon arrived at the solution. He would slaughter the cattle in Chicago, dress the beef there, and ship it to the East dressed. At first it was possible to do this only during the cold months of winter, because refrigeration was not yet adequately developed. As refrigerating cars were devised and perfected, however, the shipping of dressed beef became a year-around occupation.

It should not be supposed that Swift was alone in having these ideas. They were a contagion, a symptom of the times. Other packing plants existed in Chicago, and some of them, like Armour and Plankinton, antedated him with these very theories. In 1869 Armour went a step farther and established a plant in Kansas City to be still nearer the source of supply and save still additional freight costs. From there the packing industry, following the same general theory, extended widely in the succeeding decades— to Omaha, Denver, Fort Worth, Wichita, and elsewhere—

always with the same purpose in view. Razor-keen grew competition and Swift and Armour embarked on a new type of contest—the devising of different ways to utilize every bit of every animal. No longer were outsiders able to live well off "rubbish" which they hauled away. Not even a bird can live on the waste in a modern packing plant. Intestines, meat ends, tallow, blood—everything found some method by which it was processed. Hides, hair, bones, hoofs, horns, all became raw materials for manufacturing merchantable products. The invention of a tin-canning machine further revolutionized the industry and tinned fresh beef became an article of world export.

As meat prices went down owing to packer competition and utilization of by-products, the demand for beef and beef-products grew. Packing houses discovered that they were operating on prices so low that what formerly had been waste now constituted their actual profit, the meat itself being sold at no more than the cost of buying on the hoof and butchering. To their delight the American factory worker and bank clerk suddenly found that they could afford to eat meat, and the United States became a nation of beef eaters. At one and the same time the packing companies had evolved cheaper meat for their customers and a larger demand for the steers of the cattlemen—and a great industry, to go hand in hand with the spreading ranges of the West, had been born.

News of this metamorphosis reached down toward the South. In the summer of 1865 thousands of bronzed, gaunt, young-old men found their way back to their homes in Texas. Some of them hobbled from wounds; others were haggard with prison sickness. All had the worn look which had been stamped on their faces by four years of the bloodiest war on this continent. They were the returning veterans of the Confederate army.

At lonely ranch homes, women, mothers and wives, looked startled into thin, bearded faces which seemed to be the faces of strangers—then cried aloud in joy and thanksgiving and flew to the arms of son or husband or brother. All over Texas these scenes were repeated as families went through the experience of becoming acquainted again. As for the returning soldiers, they found things at home in a bad way. Farms and ranches were run down, sometimes devastated. The Negroes had been freed and were gone; horse stock had been taken for the use of the armies. Cultivating tools were broken or stolen, seed for planting was scarce, and money was non-existent—except Confederate money, of which there was a plenitude, the paper currency of a dead government and now not only worthless, but a painful reminder of tragic years.

One thing had not melted away in the blast-furnace of war. Even the desperate, bitter veterans were cheered by the sight of the cattle, the vast herds of cattle, which had multiplied over the years. There had been some extension of the range also. Before the War it was confined chiefly to the black land prairies, the cross timber section, the gulf coast, and the almost impenetrable brush country between the Rio Grande and the Nueces River, known as the Brasada. A line running north from Kingsville to Forth Worth would have defined, roughly, the western limit of the cattle range, and the territory to the east which it included was only about one-fourth the total area of Texas. By 1866, however, the cattle had pushed their own frontier almost to the hundredth meridian, a line roughly from Uvalde north to the corner of the Indian Territory on the Red River, near where Doan's crossing later became famous in the trail-driving days. The northern part of this western extension was not occupied yet by cattle, but its total area now extended the range to cover approximately one-half of Texas.

Nearly all the cattle in this immense area were true longhorns—products of climate, soil and surroundings, as well as of mixed bloods, evolved by Nature to fit the environment and conditions of Texas. Rattlesnakes, cactus, mesquite, dry plains, and predatory beasts held few terrors for these animals. As wild as the bison were they, but unlike the black cattle of early Texas—of which not many survived—they were capable of some degree of training as was demonstrated when the trail-driving technique was developed. How many head there were nobody knows. The census reports made no attempt accurately to enumerate them for years after the War.

Finding times difficult and discouraging when he returned home, with real privation in the cabin and a carpetbagger government in the state house, the Texas soldier turned naturally to this huge reservoir of resource. If only he could discover some means of marketing those immense herds of cattle. . . . It was inevitable that the trails should come into being.

There was, after all, nothing new in 1866 about driving steers to market. It had been done before, prior to and during the War. Texas needed one thing badly. It needed money—United States money. The thoughts of Texas, therefore, turned northward, where the money was, and where, if reports were to be credited, there was a sudden new development in Yankeeland—a springing together of the ingenuity, organization and imagination of those remote people, so strange and ununderstandable in the South, to the end that great packing houses had grown overnight and cattle were in demand. Chicago, it was understood, was the center of this demand. Railroads ran from Chicago down into Missouri, a friendly, almost Southern state. Already trails had been made to Missouri before the War—plenty of lank Texas riders had gone up them; why could not these trails be used again to reach the railheads which meant so much to the famished cow country?

As if the thought had occurred simultaneously to hundreds of cattlemen, Texas began in the spring of 1866 to gather its herds for a fabulous trek to the north. This was to be no affair of single, sporadic outfits, but a spontaneous mass movement of livestock. One thing was certain: the cattle must transport themselves to market. But Texas knew her longhorns. They were capable of incredible journeys. And Texas knew her men. The very titles many of them bore were testimony to their fiber. It seemed that every other cattleman in the state was a colonel, a major, a captain or held some other military rank fairly won in the recent war.

To this day the cattle country is liberal with its titles. A big man or a popular man quite usually is "colonel" or "judge" or even "senator" to his friends. Sometimes these honorary titles are jocularly awarded. In his *Black Range Tales*, James A. McKenna describes a frontier gathering at which five colonels were assembled. One was a legitimate officer of the late Confederate army; the second gained his title by sitting forty-eight hours in a stud poker game without rising from his chair; the third was a sort of auctioneer; the fourth was able to tell by the sound whether it was a .45 or a .45-70 Winchester that was fired; and the fifth could distinguish by the taste, with never a glance at the bottle, whether it was Old Crow or Sam Thompson or some other brand of whiskey. On the other hand, many honorary titles were tokens of respect. Such a one was that of "Colonel" William F. Cody, the famous Buffalo Bill. Another was Major Andrew Drumm, the first man to ranch in the Cherokee Strip.

Auctioneers, by virtue of their office, were colonels in the West, just as the owner of a fishing dory is a captain on the New England coast. A livestock auctioneer must be a man able to think quickly before a crowd. Usually he possesses some wit and is popular with his acquaintances. Anyone who has ever attempted to cross verbal rapiers with a colonel of the auctioneering corps will sympathize with the attorney who was attempting to cross-examine an auctioneer in a Western livestock lawsuit. Failing to shake the witness's testimony, the lawyer at length turned to sarcasm.

"They call you colonel," he sneered. "What regiment were you colonel in?"

"I reckon ye could call it the cow brigade," drawled the colonel.

"Come, come now!" rapped the lawyer. "I have asked you a legitimate question and I am entitled to a serious answer."

"Wall," explained the colonel pleasantly, "it's like this. The 'colonel' in front of my name is jest like the 'honorable' in front of yours—it doesn't mean a thing."

In Texas, right after the War, military titles presupposed actual military service, however, and the men who bore them were sufficiently hard and reckless to back up that supposition. All through the prairie country and the brush country and the cross-timbers country, they went coolly about preparing for their great gamble. They knew there were rivers to cross, Indians as a possible complication, and stampedes to be anticipated. Even when they reached Missouri there was no guarantee that they would find a market for their cattle. Yet it was speculate or starve. Texas, always good at betting on the next card, decided to speculate.

VIII

THE DEVIL IS A JAYHAWKER

Some Texas cattlemen elected to make up herds from the cattle of their own brands. Others, including a few venturesome speculators from the North, went out to buy herds for the drive. The shocking state of prices then ruling in Texas is shown by Joseph G. McCoy in his *Historic Sketches of the Cattle Trade*, which was written and published in 1874 when trail driving was at its height. McCoy tells of a personal acquaintance who looked over a herd of thirty-five hundred steers in Texas, and then struck a bargain to purchase six hundred head of his choice at six dollars a head. Then, for the next six hundred, he paid three dollars a head. In other words he obtained a herd of twelve hundred cattle, the choicest out of the original thirty-five hundred, for an average of $4.50 a head, or about forty cents a hundredweight, gross.

In that spring and early summer of 1866 Texas witnessed an awesome spectacle—cattle numbering from 225,000 to 260,000 were crossing the Red River and stringing out on their way north to a hoped-for market in Yankee-land. The herds were guided, pushed along, and carefully guarded by bronzed riders who, in many cases, had only just exchanged the cavalry saddle for the cowboy saddle, and who wore their six-shooters with a practiced ease which was evidence of long custom. Four years of war had not embittered those trail drivers. Most of them were only boys in point of years, and optimism was high among them. All Texas anticipated that the money shortage would soon be ended. How rudely that hope was to be shattered none of those first migrants dreamed.

Most of the herds took the old Sedalia trail which

crossed the Red River at Rocks Bluff ford or at Colbert's
Ferry, and angled northeast past Boggy Depot and Fort
Gibson, or else through the Kiamichi valley in the Choc-
taw country. Thence the trail extended to Baxter Springs
in the extreme southeastern corner of Kansas, then northeast to Sedalia, Missouri, at that time a railhead. Before the
War some herds had followed this route all the way to
Quincy, Illinois, and St. Louis was a thriving stocker market prior to 1860. Trail drivers on the Sedalia route sought
to avoid the Ozark Mountains to the east and yet miss the
worst river crossings.

The Texans probably anticipated a cordial welcome
when they reached their destination, but they had not
reckoned with fate, the unreasonableness of humanity, and
the instinct to prey on the helpless. First trouble came in
the Indian Territory. From the time it left the Red River
crossings the entire route passed over the lands of the Five
Civilized Nations of Indians. These tribes, the Cherokees,
Choctaws, Creeks, Chickasaws and Seminoles, were themselves farmers and stockraisers. They had not objected to
the passing of occasional trail herds through their country
before the War, but when suddenly, that summer of 1866,
more than a quarter million trampling, wild-eyed longhorns burst upon them, devouring every bit of verdure in
sight, stampeding over growing fields, and mixing up with
the Indians' stock to cause it to run away, the Five Nations
called a halt.

These were enlightened Indians, many of them with
much white blood in them, and they understood the laws
and ways of the white men. Solemn delegations met the
trail drivers on the road and with poker faces demanded a
toll of ten cents per head on all cattle passing through
their territory. That they had a legal right to do this the
Texans were uneasily aware. The Indians, furthermore,
laid out a regularly defined route and insisted that the trail
herds stay upon it, even after the grass was all eaten off. To
cap all this, the red men made the assertion—and displayed

a complete willingness to back it up with firearms—that they had the right to cut the herds at any time to search for any of their own cattle which might be mixed with the transient longhorns.

From the perspective of time, the demands of the Indians appear no more than simple justice, with terms far more liberal than white settlers ever were willing to give later, but the trail drivers fumed under them. Texans had never learned to like Indians anyway, and to talk legal technicalities now instead of debating the question with rifles was almost more than could be borne. Yet there was little anyone could do about it. Most of the trail drivers sourly paid the tax; some compounded it by giving the Indians cattle to the amount due; and a very few cocked their six-shooters ominously and rode straight ahead through the Indian country, hard-visaged and ready to shoot their way through if they were interfered with. Others, however, rather than pay or meet trouble, turned their herds east into Arkansas and skirted the Ozarks from Fort Smith northward, running the gauntlet of constant preying by outlaws who frequented those hills.

But if there was opposition to passage through the Indian country, the Texans were hardly prepared for the fierce and ruthless resistance they encountered at the southern borders of Kansas and Missouri.

One very serious disability was possessed by the Texas longhorns. They carried upon them cattle ticks—fat, blood-sucking insects with ugly, blood-bloated paunches, which fastened on a steer and sometimes hung so thickly upon him that his color and markings could not be discerned. The ticks were and are the agent which spreads the dreaded Texas or Spanish fever among Northern cattle. Immune to the disease themselves, the longhorns left behind them the loathsome parasites strewing the trails. There the ticks lay, awaiting an opportunity to crawl upon and suck the blood of any animal which passed over the trail, quickly fastening with their sharp mandibles and transferring their virus to the veins of the new host. Cold weather invariably killed the ticks, but so long as the temperature remained warm they were a deadly menace to all resident cattle in the country through which they passed.

It was not fully understood in 1866, or for some years afterward, that it was the ticks that caused Texas fever. Various theories were advanced. One group of scientists held that the disease was occasioned by "sporules," or small eggs, deposited on the grass in Texas which grew to malignant microcosms in the blood of the longhorns and were transferred to other cattle by infection. Other scientists said that Texas fever was similar to "ship fever" aboard emigrant steamships in that it was induced by hardship, lack of rest, proper food and water—which in that day were believed to be the causes of what perhaps was only typhoid fever. Even as early as that, however, the tick theory had its advocates, making a third school of thought. Nobody was sure of the cause of the disease, but of one thing the farmers of Missouri and Kansas were in unanimous agreement: in some way the Texas cattle brought the disease with them. Many of the farmers who lived in the vicinity of Baxter Springs owned dairy and other livestock. They displayed immediate and violent opposition to the entrance from the Indian Territory of Texas cattle, and enforced it with a grim line of shotguns and rifles.

This was, after all, no new thing. An item in the *Missouri Weekly Statesman*, of Columbia, Missouri, dated Friday, June 24, 1859, nearly a year before the War, reads as follows:

Texas Fever.—Three droves of Texas cattle numbering in all some two thousand, attempted to pass through our county on Saturday last, but were stopped at Grand River bridge, four miles west of this place, by committee appointed by the recent Bellmont convention. This course of proceeding may be considered illegal, but our farmers have their rights as well as the Texas drovers, and no intelligent person who is at all conversant with the past history of stock raising in this county, can for a moment blame the citizens of Missouri for adopting summary measures to protect their stock from the fearful ravages of Spanish fever.—Clinton, Missouri, *Journal.*

That very spring of 1866 the Missouri legislature had taken cognizance of the fever danger and enacted a law governing the passage of cattle from the south. The county court of Vernon County appointed several men "cattle commissioners," as did other border counties, and the office of these commissioners seems to have been to organize parties of farmers to resist the entrance of herds. Armed men were stationed on the borders of the corner counties of Missouri, particularly Jasper, Barton and Newton, in addition to Vernon, and these warlike parties were charged with the duty of turning back all "droves and herds of cattle coming from Texas or the Indian Territory."

This sort of opposition to the trail drivers, if annoying, was at least sincere and even understandable. But there was another class of borderers less wholesome. In the War between the States the Missouri-Kansas border was the focus of a bloody and destructive guerilla war in which Kansas Jayhawkers and Missouri Bushwhackers vied in deeds of violent criminality. Quantrill's raid on Lawrence, Kansas, was widely publicized, but it was no worse than the longer continued and more widespread banditry of the Kansas Jayhawkers, Jennison and Lane. To this day in certain districts of Missouri the stone chimney of a burned house, which has remained standing after the rest of the structure has fallen in ashes, is called a "Jennison monument." At the conclusion of the War, these shifty, murderous elements from both sides found themselves without congenial occupation and not infrequently joined forces in enterprises of a shady character in the southern border country of their respective states.

Agitation among the grangers against entrance of Texas herds because of the cattle fever presented these bandits, who by all were known as Jayhawkers, with an opportunity which was ideal. In the guise of "honest farmers" they attacked the trail drivers, abused and insulted the cowboys in a malignant effort to cause them to resist so that a pretext might be seized to murder them and drive away their herds. Sometimes the hot-blooded Texans went down fighting, and it is to be assumed that Jayhawkers were sometimes buried with them on such occasions. Usually, however, the cattlemen, knowing that not only their own fortunes but all the money that could be raised by all their relatives and friends in poverty-stricken Texas was invested in their cattle, were forced to swallow the abuse since otherwise they would lose everything. It was definitely demonstrated that Jayhawker gangs could be bought off at two or three dollars a head, but this was flagrant robbery and few of the trail drivers possessed the money to buy a way through for their cattle at this rate.

To be so near their goal and still be unable to reach it was heartbreaking to the Texans. The experiences of young Jim Daugherty of Denton County, Texas, just nineteen years old when he bossed a trail herd north that summer of 1866, are typical.

Daugherty was one of those who swung his herd east into Arkansas near Fort Smith to avoid paying toll to the Indians. He followed the Arkansas line north to Missouri. South of Baxter Springs he encountered several cattle herds being held up on the grass and heard from their drivers disquieting reports of the Jayhawker activities. One drover had already been killed and many cattle stolen. Uneasiness and discouragement prevailed.

Not intimidated, Daugherty left his herd south of the line with his trail crew and rode north alone to Fort Scott, where he contracted to sell his cattle to Ben Keyes. Then he returned and began to move his cattle up through the scanty timberland of southeastern Kansas. Of a sudden fifteen or twenty riders appeared out of a wooded ravine—Jayhawkers. John Dobbins, riding point for Daugherty, tried to draw his pistol and was shot out of his saddle. A blanket, flourished by a Jayhawker before the noses of the snorting longhorns, stampeded them. Up from the rear Daugherty came galloping to find what was happening. As he rode into the scowling, shifty-eyed group of marauders, he was pulled from his horse. For a time his captors debated lynching him, but his life was spared and instead the young Texan was stripped, tied to a tree, and flogged. After that he was released and the Jayhawkers rode away through the trees.

With his back bleeding, young Daugherty hunted through the woods for his friends, expecting to find them all dead and his cattle gone. To his surprise and relief the loyal cowboys had kept with the steers and rounded up all of them, except about one hundred and fifty head which definitely were lost or stolen by the Jayhawkers. Daugherty took two men, returned to the spot where the raid had occurred, buried poor John Dobbins, and placed a rough headboard over the grave. He then drove the herd back into the Neutral Strip of the Indian Territory.

Jim Daugherty had sand. That is proved by his subsequent actions. In spite of his experiences, he obtained a guide from Ben Keyes, the Fort Scott cattle buyer, waited his chance, slipped past the deadline by night driving, and sold his herd at Fort Scott. There was a saying in Missouri in the days when the notorious "Order No. 11" depopulated much of the fairest part of the state: *The Devil is a Jayhawker.* Jim Daugherty would not have disagreed with that.

Occurring again and again, attacks of this nature caused trail herds to pile up in the country south of Baxter Springs like water behind a dam. The town became the first of a long series of cowboy capitals. Some trail drivers took their cattle west past the fringe of settlement, then north across Kansas to Nebraska and eventually over the Missouri River into Iowa. Others attempted to drive east along Missouri's southern border until the hostile area was past, when the

herds were turned north toward Sedalia. But the nature of
the country was unfavorable. In the mountains and timber
those cattle which managed to reach Sedalia became so
thin and footsore from the rocky trails that they sold for
less than enough to pay expenses.

More than half the trail herds were held up at Baxter
Springs in the hope that conditions would change and
permit them to go through the blockade. Autumn came
on with cold weather and the cattle still were there. The
frost killed the grass, which, unlike the buffalo and grama
grasses of Texas and the Western plains, dried up when
dead and lost its nourishment. Helplessly the trail drivers
watched their herds starving on their feet. To add a climax
to misery, somebody—perhaps some of the Jayhawkers—set
fire to the prairie and destroyed much of the pasture that
was left, poor as it was. In despair many Texans sold out
at ruinous prices. Others were defrauded by glib-tongued
strangers through bogus checks and drafts. The country,
filled with blockaded cattle, became a focus for swindlers,
crooks and thieves of all kinds. Some herds were stolen
outright. Many cattle died. Of the more than a quarter
million Texas longhorns which had been brought through
peril and with infinite labor to the Kansas border that
summer of 1866, very few were sold at anything approaching a profit.

As has been said, while cowboys camped near it, Baxter
Springs was briefly wild. Gambling halls and saloons flourished. A "red light" district sprang up and the hard-eyed
trulls of the camp beckoned from their windows. It was
a brief, evanescent capital of sin, but it was the first of a
scrofulous series, including Abilene, Ellsworth, Newton,
Wichita, Dodge City and many another.

George C. Duffield, who made the drive of 1866, was
a young Iowan. He owns one great distinction among the
trail riders—he alone of all of them kept a diary. The document, published in the *Annals of Iowa* in 1924, gives an

illuminating and intensely human view of the difficulties of the trail. With Harvey Ray, his partner, Duffield went down the Mississippi by steamboat, and to Galveston by coasting steamer, then overland to the Colorado River country where they bought cheap cattle and made up a herd to be returned to Iowa. It was April 29, 1866, that the northward march began.

Duffield's journal gives a picture of the ensuing hard and exasperating journey. Stampedes occurred May 1 and May 6—and pretty continuously thereafter. Each time days were wasted trying to find the animals that were lost. By May 9 the young Iowan wished fervently he was through with his task, as his entry in his diary shows: "Still dark & Gloomy River up everything looks *Blue* to me." Four days later another maddening stampede during a thunder storm added to his gloom, although he recovered all but fifty of his steers: "all tired Everything discouraging."

But when he reached the Brazos the real trouble began. His approximately one thousand cattle were divided into three herds, with twenty cowboys as trailers, and it took three days to make the crossing. Cattle and horses were swum across and provisions and camp equipment were "rafted" over. Unfortunately most of the "Kitchen furniture such as camp Kittles Coffee Pots Cups Plates &c &c" were lost in the process. After rounding up the cattle on the other side of the river, "all Hands gave the Brazos one good harty dam," and rode away without joy.

Rain fell and the wind blew almost constantly on the journey and the Texas cowboys with the herd grew sulky. Some of them quit. On May 20 Duffield wrote: "Rain poured down for two hours Ground flooded Creeks up —Hands leaving Gloomey times as ever I saw."

Most of their few remaining cooking utensils were lost in the crossing of the Trinity, and the following night, May 23, "Hard rain that night & cattle behaved very bad —ran all night—was on my Horse the whole night & it

raining hard. Glad to see Morning come counted & found
we had lost none for the first time—feel very bad."

Three days were required by the dolorous Mr. Duffield
to put his herd across the Red River and at that crossing
the first tragedy of the journey occurred. A cowboy named
Carr, caught in the swirl of the tide while working with
the swimming herd, was drowned. To signalize their passage of the river, the perverse longhorns stampeded again
the following night. Next day the diary noted: "Hunt
cattle again Men all tired & want to leave. am in the Indian
country am annoyed by them believe they scare the Cattle
to get pay to collect them. . . . Two men and Bunch
Beeves lost—Horses all give out & Men refused to do anything." And on the succeeding day: "Hard rain & wind
Storm Beeves ran & had to be on Horse back all night.
Awful night. wet all night clear bright morning. Men
still lost quit the Beeves and go Hunting Men is the word
—4 p.m. Found our men with Indian guide & 195 Beeves
14 miles from camp. allmost starved not having had a bite
to eat for 60 hours got to camp about 12 m *Tired*."

For several days things went a little better, although
the country was boggy with the heavy rains and the rivers
and creeks gave constant trouble. But on June 12 there is
the following entry: "Hard Rain & Wind Big stampede
& here we are among the Indians with 150 head of Cattle
gone hunted all day & the Rain pouring down with but
poor success Dark days are these to me Nothing but Bread
& Coffee Hands all Growling & Swearing."

It was enough to make them swear, but by no means
were their troubles over. On June 17 they reached the
Arkansas where Duffield spent four more days swimming
his depleted herd across that river which was a raging torrent, roaring in spate, owing to the heavy rains. "Worked
all day hard in the River trying to make the Beeves swim
& did not get one over." the mournful young chronicler
of the trail wrote at the end of the first day's efforts at
the Arkansas. "Had to go back to the Prairie Sick & dis-

couraged. Have *not* got the *Blues* but am in *Hel of a fix*."

Eventually, however, the cattle were crossed and the herd reached the vicinity of Baxter Springs July 10 with no further losses. There Duffield found the cap and climax to his woes—the grangers and Jayhawkers were in charge of the border and the cattle could not pass through Missouri.

Several days were spent in fruitless scouting and negotiations. In spite of Duffield's lugubrious moans on the trail, he seems to have had plenty of decision and nerve, and he showed at this crisis more enterprise than most of the Texans—possibly because he knew the country in which he now was better than they did. His entry of July 25 reveals his decision: "We left the Beefe Road [trail] & started due west across the wide Prairie in the Indian Nation to try to go around Kansas & strike Iowa. I have 490 Beeves."

It was a wise decision. Swinging his herd to the west he passed around the settlements north to the Nebraska line. One cannot but sympathize with the young trail driver in his woes and even after the passage of sixty years there is joy in knowing that at last he came to the end of the sorrowful road. Early in September, the ancient journal records, he reached the Missouri River near Nebraska City and ferried his few hundred remaining cattle over into the promised land of Iowa.

There is an expression still current in the American language: "In spite of hell and high water."

It is a legacy of the cattle trail, when the cowboys drove their horn-spiked masses of longhorns through high water at every river and continuous hell between, in their unalterable determination to reach the end of the trail which was their goal.

IX

MEN WHO DIDN'T SCARE

While most of the Texas cattlemen were piling up their herds on the frontier at Baxter Springs and watching their cattle starve, or were being defrauded and robbed by the Jayhawkers, two trail-driving exploits were performed on far-distant frontiers which, while not partaking of the mass importance of the cattle concentration in the Indian Territory, by all other measurements rank with it. These exploits occurred two thousand miles apart and an equal distance from Baxter Springs.

A wandering halfbreed named François Findlay, trapping in western Montana, stumbled, in 1852, on a gold-bearing placer in a creek which flowed into Hellgate River, halfway between where now stand Garrison and Drummond, Montana. The stream, known as Gold Creek, started prospectors roaming all over the Montana mountains, a restless, floating population, always on the *qui vive* for new rumors of "strikes." In the early '60's strikes occurred with spectacular frequency in Nevada, Idaho and Montana and the gold fever consequently mounted in intensity. At Elk City, Oro Fino, Florence and Warner Creek in Idaho, gold fields opened just as the War between the States broke out and seven thousand men rushed into the Northwest.

Not, however, until the fabulously rich discoveries were made on Grasshopper Creek in southwestern Montana in 1862 did the real delirium come. Three thousand prospectors made a headlong charge into those "diggings" almost overnight, and Bannack, the first territorial capital of an area which hitherto had been almost entirely Indian country, rose in lurid, mushroom growth. The very next year came the great strike at Alder Gulch and by 1864 Virginia City had a population of ten thousand and had wrested from Bannack, fifty miles west, the territorial capital. But Virginia City did not long enjoy the victory. Last Chance Gulch came into being to the north and Helena sprang up to become the capital, which it remains today.

By the end of 1865 the gold country of Montana and Idaho had acquired a heavy population of typical mining people, and the problem of feeding them became acute. Few of the newcomers gave any thought to producing food; their interest was focused on the yellow metal they mined or sought. A similar situation had existed in California when the first gold madness struck and farmers, *vaqueros*, merchants and others left their occupations to rush to the diggings. Even seamen deserted their ships, which were left at the docks without crews. A biographer has said that P. D. Armour's experiences in California, when he observed the effects of a lack of food supply in a population, inspired him with the idea of a great packing enterprise to serve large numbers of consumers on a mass production basis.

A similar inspiration came to a man named Nelson Story in Montana, when he visualized the possibilities of bringing cheap Texas beef to a market where the demand was acute. The result was one of the most extraordinary ventures in the history of the West.

Nelson Story was a boy fresh from Ohio when he arrived in the West, reaching Fort Leavenworth in 1856 with thirty-six dollars in his pockets. At first he freighted, plying between Leavenworth and Denver and gradually working up until he owned his own outfit of wagons and oxen. The excitement of 1863 took him out into the new gold fields of Idaho and Montana and he was one of the first men in Alder Gulch, where he became a captain of the vigilantes who broke up the Plummer gang of road agents—Westernism for stage robbers—by hanging twenty-two of its members. In the early months of 1866 he "struck it rich," taking $30,000 out of a placer claim near Summit at the head of Alder Gulch.

Washing gold, however, was too monotonous for Nels Story. He turned the claim over to two brothers, shipped his gold to Kountze Brothers, in New York, and received in exchange $40,000 in greenbacks, currency being at the time depreciated because of the War. Off the top of this he peeled $10,000 and enough more for expenses and sewed the money in his clothes, put the remainder in safe keeping, and with two young men, Bill Petty and Thomas Allen, set out for Fort Worth, Texas. His original object was to buy cattle which, he heard, were selling dirt cheap in Texas, and transport them north to the market which he envisioned as opening up at Kansas City.

At Fort Worth Story bought one thousand longhorns at ten dollars a head and, hiring a crew of cowboys, started the herd across the Indian Nations. As we have seen the year 1866 was wet, and Story swam turbulent rivers, dodged Indian toll collectors, and evaded thieves, both red and white, besides heading off stampedes all the way. Nobody was killed in his outfit, which was a wonder, and few animals were lost.

But when the Kansas-Missouri border was reached at Baxter Springs, Story found the same discouraging situation awaiting him which confronted all the other outfits there—the grangers and Jayhawkers held the border. Story made a quick decision. Ten thousand miners were in and around Virginia City, and other thousands in the near-by mountains—and he remembered that beef was very scarce there. If he could reach Montana with his herd he would, in the language of the day, "make a killing."

Out of the jam below Baxter Springs he took his steers and headed them west to go around the fringes of the Kansas settlements. North the cattle turned, crossing the Smoky Hill River, then veered east along the Kaw through Topeka to Fort Leavenworth. There Story bought a wagon train, loaded it with groceries and provisions, hired experienced bull whackers, and purchased work oxen. Out of the fort he presently took the Oregon trail with his train of provisions and his herd of Texas steers, his face set westward for a tremendous gamble in which his life was at stake with his fortune.

When the caravan crawled up the Platte valley to Fort Laramie, Wyoming, army officers shook their heads. The entire Powder River country through which Story must pass if he followed the Bozeman trail, the most feasible route to Montana, was swarming with hostile Sioux, said the soldiers. Colonel H. B. Carrington had just started his luckless Powder River expedition. He had built Fort Reno on the Powder and was erecting Fort Phil Kearny on the Piney, with Fort C. F. Smith still to be constructed farther north. The series of events which was to culminate the following December in the annihilation by the Sioux of

Captain William J. Fetterman's command was well under way. To illustrate the peril of the country in the days when Story proposed to cross it, more than one hundred and fifty men were killed around Fort Phil Kearny in the six months after work was begun on the construction of the post, while the Fetterman disaster was the worst defeat suffered in Indian fighting by the American army in the West until the Custer debacle of 1876.

Story heard the warnings and ignored them. He had twenty-seven men and he armed them all with Remington breech-loaders, rapid-fire rifles which were new to the time and country. Then, early one morning, he resolutely headed his cattle northward. The thing which the officers had predicted was not long delayed—near Fort Reno a band of Sioux charged over a rise, knocked down two cowpunchers with arrows, and swept away with part of the cattle herd.

Years later John B. Catlin, one of Story's riders, told the aftermath. "How many cattle did you lose?" he was asked.

"Not a single head. We just followed those Indians into the Bad Lands and got the cattle back."

"Did they yield the steers willingly?"

"Well, you might say so. We surprised them in their camp and they weren't in shape to protest much against our taking back the cattle."

His wounded men Story left at Fort Reno and went on. But at Fort Phil Kearny he found fussy, indecisive Colonel Carrington ready to forbid any further progress northward. Carrington was afraid of the Sioux and he thought everyone else should be afraid also. It was certain death, he told Story, to progress farther because the Indians, who were dangerous enough south of Fort Phil Kearny, were ten times worse to the north. Carrington ordered Story to halt his herd there and wait for permission to proceed. The trail drivers built two corrals, one for the beef steers and one for the work oxen, three miles from the fort, which was as near as they were permitted to come because Carrington said all grass near the post was needed for government animals. So far away were the cowmen that they could count on no help from the fort, and once they actually had to stand off an attack with the post almost in view.

It was an impossible situation; Story felt that destruction was certain if he delayed longer. One night he called together his men and took a vote on whether they should slip away and continue their march past the soldier guard. Every man, except George Dow, favored the action. Story promptly threw a six-shooter on Dow, put him under arrest so that he could not inform the soldiers, and that night the party slipped the cattle around the fort and disappeared into the Indian country. When they were a day along the trail Story offered to free Dow to return to the fort, but the man was afraid to go back to Phil Kearny alone and chose to continue with the party. On this part of the drive Story adopted a new policy. He trailed the herd only at night. In the daytime the cattle were herd grazed under guard. Day after day the daylight grazing continued; night after night the slow, plodding pace was followed.

Twice the Indians attacked the crawling procession. Near Clark's Fork of the Yellowstone, a shot far ahead attracted Story's attention. One of his men was hunting for the party, about a mile ahead of the herd. Story rode forward and mounted a rise just in time to see a band of fifteen or twenty Indians reach the hunter, who had been shot down. Two of the savages rode up, one on each side of the prostrate man, lifted him between them, and the whole band galloped off to the top of a hill half a mile away. With a group of his cowboys Story went to the rescue. The Indians disappeared, but the corpse of the hunter was found on the hill. He had been scalped and his body lay nailed to the ground with arrows. That was the only death. "There were only twenty-seven men in our party,"

Catlin recalled. "There were about three hundred troopers at Fort Phil Kearny. But the Indians were worse scared of us with our Remingtons than they were of the troopers with their Springfield army guns. After we got 'em scared it was easy for our twenty-seven to stand off three thousand reds with their bows and arrows."

Story's outfit left Fort Phil Kearny on the night of October 22. On December 9, 1866, the entire herd and train-load of groceries rolled into Virginia City, where Story was greeted at the head of the trail by his young wife whom he had left there. In more ways than one this was a record-making trip. After Story's outfit went up the trail it closed behind him as if it had been barred, with Indians always on the watch, malignantly ready to kill any who went up it. Not for four more years would another trail herd pass up the Powder River valley.

The apotheosis of the cowman was Charles Goodnight. He led the way in so many things that his name is tied up with almost every turn of Southwestern history. He had been a freighter, bull-whacking between Fort Worth and Palo Pinto County before the War between the States, and in so doing froze one of his feet so that he limped ever after. During the War he served four years with the Texas rangers, fighting Indians, Mexicans, renegade cow thieves and road agents. He it was who saved the life of Cynthia Ann Parker when she was captured in 1861. Goodnight was riding beside Captain Sul Ross when the latter aimed at a fleeing Indian woman. The wind blew aside her blanket and Goodnight shouted to Ross to hold his fire, that the woman was white. She turned out to be the lost Cynthia Ann.

The Comanches took advantage of the War to handle savagely the northwestern frontier of Texas. Entire counties were depopulated. Nevertheless, when peace came and his service with the rangers ended, Goodnight went out into Palo Pinto County where vast numbers of cattle swarmed, unowned and unbranded. In partnership with his step-brother, J. W. Sheek, he established a ranch on that remote prairie, bought the C V brand and began to build up a herd in defiance of the Indians. After indescribable dangers and labors the partners accumulated seven thousand cattle; then Goodnight left Sheek to watch the Palo Pinto range while he went still farther out into Comanche territory, where Throckmorton County now is. With him he took three thousand cattle and loose-herded them. One night Mexicans and Indians ran off two thousand head and Goodnight with his small force of hands had to watch helplessly.

When Texas stirred for the giant effort in 1866, Good-night did some canny thinking. Other cattlemen were driv-ing north. Goodnight knew that to the west lay another rich cattle market, if he could only reach New Mexico. General Carleton's long campaign against the Apaches had accomplished some results, and Kit Carson had made his famous march down Canyon de Chelly, rounding up the Navajos as he went. Between them these soldiers collected some seven thousand Indians on the Bosque Redondo reser-vation in the Pecos Valley near Fort Sumner. The govern-ment had to ration these Indians as well as its soldiers, and beef was scarce and costly in New Mexico, which was primarily a sheep country.

Great difficulties must be surmounted if Goodnight was to reach that market. Between him and New Mexico lay the most dangerous Indian country in the Southwest. It was ridiculous to think of trailing a herd from the upper Brazos to Santa Fé—the Comanches would have gobbled the outfit, steers, horses, men and all. A possible route did, however, exist. It lay southwest across the Staked Plains, following the old stage line called the Horsehead route, established in 1846 by General Pope. The country through which it ran was so far south and so desolate that the Comanches usually avoided it. One great obstacle to this trail had to be considered—a terrible, ninety-six mile stretch from the Middle Concho to the Horsehead crossing on the Pecos, over which not a drop of water existed except in the rare and unpredictable periods of desert rainfall. Experienced cattlemen said the Horsehead route was impossible of passage by cattle, but Goodnight, a man of great imagination and daring, believed he knew the qualities of Texas longhorns well enough to risk it.

By a fortunate circumstance he one day met Oliver Loving, at the time generally considered the most experienced trail driver in Texas. As early as 1858 Loving had trailed a herd north through the Indian Nations, across Missouri to Quincy, Illinois, establishing the Sedalia trail. Two years later, in August, 1860, while the gold rush was on in Colorado, he pointed a herd of one thousand steers north toward the Arkansas River with the idea of finding the mountain market. The drive began August 29, and Loving, aided by John Dawson, who owned an interest in the herd, trailed the cattle north, mile after weary mile, skirting the western limits of Kansas settlement, and finally reaching the Arkansas at about the mouth of Walnut Creek in the Great Bend. From there he followed the river west to Pueblo where he wintered his cattle before disposing of them in Denver. During the War Loving was busy driving cattle to the Confederate armies, and the final defeat of the South left him almost penniless.

Loving was in his fifties when Goodnight proposed his Horsehead plan, while Goodnight was just thirty. The elder man at first warned of the danger involved in the proposed route, and then, his imagination taking fire at the boldness of the proposal, asked to be permitted to share its risks. That was exactly what Goodnight wanted. A historic partnership was formed. The two gathered a herd—two thousand head of mixed stuff, cows, steers, and bulls, even calves. It has been said that the first chuckwagon in all history was used on this trip, the idea being Goodnight's. He built a huge vehicle of the yellow wood of the *bois d'arc* or Osage orange, equipped it with water barrels, and yoked to it ten pairs of oxen. Confederate army veterans were picked for trail hands as a precaution against the Comanches, in spite of the fact that the route chosen was not much frequented by the Indians. In June the drive began.

Trouble started early. Cows find it difficult to keep pace with adult steers, and young calves cannot keep up at all. In the beginning Goodnight attempted to carry along in the huge chuckwagon the newly dropped calves, until they acquired strength to follow their mothers, but the experiment failed. He was compelled to kill these innocents. It was a thing he hated and he resolved never again to drive a mixed herd across the Horsehead route. Southwest the herd rolled slowly, from the upper Brazos through Buffalo Gap and then up the Middle Concho, forty miles to its head. There the cattle were permitted to rest and were encouraged to fill their skins with water, in preparation for the terrible test they were about to undergo.

The dreaded ninety-six mile dry leg of the drive began. Far out in the Staked Plains from the Rio Concho's headwaters was Centralia Draw, down which had gone the Butterfield stage lines to the Horsehead crossing before the War. From the plateau the trail debouched through the Castle Mountains, dropping down into the Pecos valley through Castle Gap, twelve miles from water.

All day the herd toiled across the high plains, with the bitter alkali dust rising around it and the men riding masked to the eyes with their neck handkerchiefs so that they could breathe in the stifling cloud. The second day the cattle were very restless with thirst, and that night they milled and trampled about through the dark hours, so that nearly everyone in the outfit was called out to hold them. Next morning Goodnight spoke to his partner:

"Loving, this will never do. Those cattle walked enough last night to have got to the Pecos. Camping won't work; we've got to let them travel."

"Guess you're right, Charlie," said Loving. "Take charge of 'em and see what you can do."

All the next long day the herd tramped on. Haggard cowboys, their lips cracked and their eyes bloodshot, kept the cattle moving. The water barrels were emptied and the ten yoke of oxen dragged along a dry chuckwagon. The stronger cattle forged ahead and the weaker lagged behind until keeping the herd together became a worrying problem. Oliver Loving took charge of the drag, to save as many laggards as possible, while Charles Goodnight, on a tough, blocky horse, rode here and there, now at point, now on the flank, directing and encouraging. The point riders fought to hold back the leaders of the herd so they would not be separated too far from the drag. At last Goodnight fastened an ox-bell to one of the point riders' horses, and when the drag could no longer hear that bell a cowboy was sent forward to call for the lead cattle to be held until the rest of the herd came up.

By the third evening animals and men alike were nearly crazed. The cattle formed a pitiable spectacle. Their ribs stood out like basketwork, flanks sagged gauntly, tongues lolled from mouths and heads hung low. Above the yells and whistles of the cowboys rose the bawling, bellowing, and moaning of the suffering herd. It was a sound never to be forgotten, but there was only one thing to do—keep the cattle moving. If they ever stopped and lay down they would never rise again. On and on the suffering drive forged. The fourth morning they entered Castle Gap and were greeted by a cool breeze which sent the cattle into a near stampede under the belief they smelled water. As they emerged from the gap, the Pecos was seen, twelve miles away.

"When the cattle reached the water they had no sense at all," said Goodnight. "They stampeded into the stream, swam right across it, and then doubled back before they stopped to drink."

Proof of the peerless endurance of the longhorns is the

fact that although this trail herd was a mixture of ages, sexes, weak and poor cattle, Goodnight and Loving lost only three hundred head on the terrific drive. The animals were rested for a time on the Pecos, then driven north to Fort Sumner. Their arrival occasioned profound rejoicing among seven thousand Indians who were almost starving. Beef was so scarce that it was selling at sixteen cents a pound, dressed. The partners closed out their beef steers at eight cents a pound on the hoof—a price unheard of for Texas cattle in that day.

At Fort Sumner Loving and Goodnight separated. They had made history with their successful drive. As soon as it became known that the Horsehead route was negotiable by cattle, it would be flooded with herds and prices would fall in New Mexico. It was decided that while Loving took the stock cows and bulls not sold as beef at Fort Sumner north to Colorado, Goodnight should hasten back to Texas and bring another herd across.

The trail Loving made from Fort Sumner to Denver became known as the Goodnight-Loving trail. On the way he crossed the Raton Pass in southern Colorado, where he met a cantankerous old mountain man, Uncle Dick Wootten, who had built a road across the pass, and now sat guarding it with his eight-square Hawkens rifle, collecting toll from all who used it. Loving was forced to pay ten cents a head for his cattle. It made the Texan indignant, but he sold in Denver at a good profit and that was balm to his feelings.

Goodnight returned to Texas with twelve thousand dollars in gold packed on a mule's back. One night the mule stampeded and almost lost the profits of the drive. But Goodnight caught him. Reaching Texas the cattleman quickly gathered a second herd, paying gold this time and able therefore to pick his steers. Because of the superior grade of cattle he lost only five head on that trip over the Horsehead, but he reached the Pecos so late that he had to winter in New Mexico before disposing of his herd.

The partners had established a great cattle trail and set a trend in the distribution of Texas longhorns. The next year was to see many drovers follow them and disaster overtake one of the partners. Because of its audacity and the triumph over hardships and dangers, the Horsehead drive of 1866 rivals Nelson Story's epic trailing through Red Cloud's Powder River country the same year. With W. D. H. Saunders' war-time drive across the Mississippi River, these are to me the most remarkable of the trail drives—certainly the most interesting.

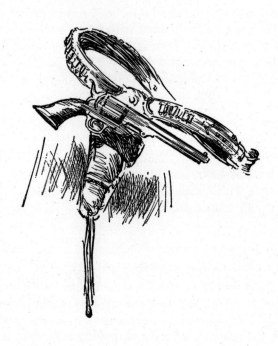

X

THE BRUSH POPPERS

Although the drive to Missouri in 1866 had been a stunning disappointment, the very fact that some cattle were sold and that a few, a very few, of the drovers had made a profit, sang through Texas like a high note of hope. When the spring of 1867 opened activity began again in the cattle country. For the next several years gathering of herds for the trail was to assume the aspect of a major activity in the entire state.

The cattle industry had many facets, each of them distinctive. Ranching was one, and it varied according to whether it was in brush country, desert, coastal regions, plains or mountains. The roundup was a distinctive institution, and it too varied with the locality. Throughout Texas the brush made a heavy screening, sometimes so tangled with thorny growth that it was virtually impossible to force a way through it. In the *Llano* hills were splintering canebrakes; east of the San Antonio River was a great stretch of blackjack thicket; on the upper Rio Grande were tangles of *tornillo*, mesquite and chaparral; and between the Rio Grande and the Nueces was the *Brasada*, composed of the worst jungle in the entire cow country of America, of combined varieties of plants, all armed with dirk-like thorns. Through all these growths the cattle tunneled paths very much as hippopotami tunnel paths through the tall sedge by the rivers of Africa. To the cattle the paths were known intimately, including all the turnings and forkings, and for a man on a horse to attempt to run the animals down by following their trails was next to impossible.

Ways existed, however, to catch the wild longhorns.

These preferred to rest in the daytime in their dark jungles, coming forth at night to feed. Knowing their habits, the *vaqueros* often employed a decoy herd and a corral. The decoys were tame, branded cattle—tame, that is, according to Texas standards. The tame cattle were taken to a place near a corral which had been built in the brush, with wide-spreading wings leading to the gate. Now the cowboys made a wide circle, seeking to surround as great an area as possible. All this was done with the utmost care to avoid undue disturbance. When the surround was complete, the riders began to work very quietly through the tangle toward the corral and the decoys. It was exciting work, with every cowboy and *vaquero* keyed to the limit of tension, because it was the duty of all and a point of honor to prevent any cattle from breaking back through the advancing cordon.

Of a sudden there would be a snort and a crash up ahead, followed almost immediately by the pounding of hoofs. The wild cattle had smelled or seen one of the riders—and stampeded. As they thundered down their crooked paths, picking up other cattle in their flight, they were followed by the riders, who risked loss of eyes, broken limbs, even death, should they be raked from their saddles by the clutching thorny branches as they rode low along their horses' necks.

Now came the usefulness of the decoy herd. As the longhorns stampeded up to the tame cattle and found they disinclined to bolt, the wild herd slowed down and turned back. Presently the cowboys, riding quietly once more, could see their tame cattle mixed with the wild ones. Now the leader, or *caporal*, began singing a cowboy ballad some distance away. Gently the cowboys urged the herd in the direction of the sound and for a wonder the cattle usually followed docilely after the unseen voice until the *caporal* led them into the wings of the trap and the horsemen, following, drove them into the corral.

Once corralled, the cattle were left without food or

water for two or three days, at the end of which time they were much tamed and so preoccupied with the desire to drink that when released they did not at once stampede off into the brush. Meantime the cowboys roped and branded all the mavericks and it was usually possible to drive them to where some trail herd was in formation. Individual animals which proved intractable were "necked" to tame cattle—a process less gentle than it sounds, consisting of tying the two together by their necks with a rope. The wild animal might buck and try to run; he might even communicate his intentions to the tame one; but cattle rarely decide to run in the same direction at the same time, and after the two had fought it out at cross purposes for a day or so, both would be exhausted and ready to fall in with the appointed scheme of things.

At other times cowboys and Mexican *vaqueros* rode into the brush and roped and hog-tied cattle right in the thickets. A very short lariat was used in this work—rarely more than thirty feet—since to use one any longer would have been impossible in the twisting paths through the dense growth. To rope and throw a full-grown steer in the middle of a tangle of thorny chaparral, mesquite, and cactus, to hog-tie, brand and ear-mark him, is a feat both difficult and dangerous. Yet plenty of cowboys in Texas performed that feat not once but many times a day, and day after day, as an ordinary part of their daily work.

Many public corrals were scattered through the Texas brush country at this period—structures built by co-operative effort among the cattlemen. They were made by planting poles deeply in the ground and lashing them together with green rawhide which dried to a consistency almost like iron.

If wild cattle were too wary to be caught in the ordinary ways, cowboys sometimes resorted to night hunting. After sunset, when the moon was full and there were no clouds to obscure the visibility, the cow hunters hid in the brush near one of the little open prairies on which cattle came out to graze in the darkness. As the animals filed forth, each cowboy selected one as a quarry and at a signal all dashed forward, nooses swinging. Rope work by moonlight was tricky, but these cowboys fastened on and threw their cattle more often than they missed. Occasionally a *vaquero* would "tail" a cow or steer—that is, he would ride past the animal, stooping as he did so to seize the end of its tail, twist it a time or two around his saddle horn, and thus whirl the creature around, throwing it so heavily that it usually lay still until he could hog-tie and brand it. The feat required the absolute pinnacle of horsemanship and co-ordination. Yet it was not uncommon.

Undertaking to drive one to three thousand staring-eyed and snorting longhorns even out of the brush in any given direction is seen to partake of adventure. Much more so is this true of trailing a herd on a continuous and planned route of several hundred miles, over brimming rivers, in storm and drouth, for weeks and even months at a time. It was necessary, however, to drive the cattle, and the cow-knowledge of the old *vaqueros* of Spanish days had been so well taught to the more modern white cowboys of Texas, that long trail drives were undertaken without concern—as a matter of routine. Trail driving, as a matter of fact, became a definite profession quite apart from other cattle work. You hear of men who made five, six, ten, twelve trips up the trails, and since the entire history of trail driving was short, and since one drive in a season was about all a man could expect to accomplish, that means these individuals spent their lives pretty constantly at trail driving during the great cattle migration.

There was good profit in trail driving—if a man was lucky. Of course if he lost his herd by stampede, or theft, or blizzard, or flood, or struck a panic market, or had any one of a hundred other things befall him which continually threatened, his luck was bad and he not only made no profit, but lost his work and investment and perhaps

his life as well. Colonel Ike T. Pryor, one of the great
cattlemen, who died very recently, took the trouble to
cast up figures on each of fifteen herds he drove from
southern Texas to Wyoming and Montana in 1884. Then
he averaged them and was able to arrive at a fairly close
estimate of the normal cost of operation.

Pryor found that three thousand cattle comprised the
most manageable and economical herd to handle. For such
a herd ten cowboys, including the trail boss, were required,
and each man was furnished six horses. In addition there
was a cook who drove the chuckwagon, with food and
bedrolls for the men, and prepared the meals. The trail
boss received about $100 a month; the men, including the
cook, $30 a month each. That was an outlay in wages of
$400 a month, and with an additional $100 a month as
the estimated cost of provisions, a total of $500 a month
was required to move a trail herd of three thousand cattle.
The cattle would travel four hundred and fifty to five
hundred miles a month, which figures out about a dollar
a mile for the entire herd—sufficiently cheap when compared to modern costs of transportation. Pryor computed
that his average expense of transporting a herd of three
thousand cattle from southern Texas to Montana, a distance of three thousand miles, was about $3,000, which is
another way of arriving at the dollar-a-mile average, and
works out to a dollar a head for driving the cattle that
distance.

Thus a cattleman who paid, say, $8 a head in Texas for
his steers and later sold them at $20 in Kansas, had a wide
margin to work on. He could charge off a fair percentage
in losses from strays, stolen stock, and deaths on the road,
deduct his trailing expenses, and still make thousands of
dollars on a drive—provided he did not, instead, lose everything on the way up.

A great, leaping excitement attended the start of every
trail drive. The cattle, frequently contracted for in small

bunches, were brought together, perhaps in a large corral. Small herds were the rule in the earlier years. One thousand head in the first days of the trails was considered about the maximum manageable number. Later the herds increased in size until in the middle '80's, as Ike Pryor showed, three thousand head was considered about right, and some bunches of ten thousand were drifted through. There was even one herd of twenty thousand driven from the Pecos across the Panhandle into the Indian Nations— but there were special circumstances surrounding that drive which will be dealt with later.

After the herd was gathered the trail boss lined out his drivers. He hired a cook, who not only had to be proficient according to the rather unexacting culinary standards of the cow country, but also had to be an expert wagon driver—"bull whacker" or "mule skinner" according to the type of animals hauling the chuckwagon. Many old-timers have paid tribute of varying kinds to the trail cook. Emerson Hough spoke feelingly on this score. The cook might be a Negro, a Spaniard, or a "Portuguee," he said, and it was almost certain that he would be unlovely of countenance and with perhaps a notch or two on the haft of his knife. The boys on the drive very likely would call him "the old lady," but they would do this beyond earshot of him, not only because he was a dangerous fighter, but because it did not pay to incur his ill will, the luckless waddy who did so always receiving the poorest victuals at meal time.

A strange and wonderful device was the chuckwagon, the original of which, as we have seen, was used by Charles Goodnight on the first Horsehead drive. As developed and standardized, it was either a good stout wagon, or in the early days a two-wheeled *carreta*, covered with a canvas tilt. Provisions, bedding for the cowboys, slickers, extra clothing, and other personal belongings were stowed in the wagon bed. A large water barrel was securely fastened toward the front, with a spigot which ran through the

side of the wagon, permitting water to be drawn from
the outside. On more than one occasion that water barrel
was a life-saver. In a pinch its supply could be made to
last two or three days for all hands. At the rear of the
wagon was the chuck box, with dishes and cooking utensils. Dishes were tin or iron ware. The lid of the chuck
box possessed a folding leg so that when it was lowered
it formed a table on which the cook could work. Below
the wagon was slung a peculiar device known as a
"cooney" or "caboose." It was a cowhide, fastened under
the wagon by its corners and forming a loose sack into
which were thrown pieces of dry wood, dry cow chips,
and other fuel for use in the treeless areas. An ax, hammer,
hatchet, spade and other simple tools were stored in a
box in front. The presence of the spade should be noted.
It was there for a purpose. You cannot dig a grave with
an ax.

With his outfit ready, the trail boss next "trail branded"
his herd. Composed as it was of many different brands,
it was necessary to have some uniform designation.
Usually, in the early days at least, the herds were nondescript collections of animals of both sexes and all ages
and conditions. This was because of the free and open-
handed tradition of Texas. Cattle were purchased by the
head instead of by the pound, and all mature cattle were
listed as "beeves" in that early period. The drover who
refused to accept animals on the ground they did not
meet his specifications was a "short sport" and a "skin-
flint" according to Texas notions. They said he "would
skin a flea for its hide and taller." So strong was the prejudice on this score that no Texan would incur the stigma,
and even Northern buyers learned the power of public
opinion and accepted the custom.

It was the rule for the several owners of cattle making
up a trail herd to give bills of sale to the drover which
set forth the brand, number, earmarks and other identifica-
tions of each contribution to the herd. Particularly was this

necessary when trail driving had been in progress for a few years and state inspectors began to scrutinize each traveling bunch of cattle.

Some of the trailing operations were on a very large scale. Ike Pryor, in the year quoted, drove forty-five thousand cattle up the trail. He employed a minimum of one hundred and sixty-five men and one thousand horses in the enterprise; the cost of the movement was around $45,-000, and an investment of more than a million dollars was represented in the cattle alone, to say nothing of the horses and equipment. Pryor's herds went north almost in sight and hearing of one another, "like sections of a large train in modern railroading," as someone has put it, with Pryor himself accompanying the last herd to see that the strays from those preceding were picked up.

But Bill Jackman, another trail driver, found at least one steer that the vigilant Pryor had overlooked. It had the Pryor road brand and, following custom, Jackman brought it along with his own cattle. Trail drivers all did that. Careful account of such animals was kept and when they were sold it was a point of honor to "settle up" with the owners.

On this occasion, however, shortly after the Red River was crossed, a band of Indians rode up to Jackman's herd. There were thirty or forty of them, bucks and squaws, and they began the usual loud demand for "wo-haw" and "chuck-a-way," those being the red man's expressions for "beef" and "food." Parenthetically, "wo-haw" was derived from the language of the ox team bullwhackers in the Indian Territory. Commands to oxen were "Gee" (right), "Haw" (left), and "Whoa" (stop). The sound of the last two appeared to possess charm for the aboriginal ear, and combined they formed the word "wo-haw," which was a generic term for cattle, beef, or anything related to them.

When Indians appeared it was ordinarily considered good policy to cut out a steer or two from the drag, where the lame, lazy, and sickly cattle were, and present them to the tribesmen. The leader of this band, however, showed Jackman a piece of very thumb-marked and grubby paper, on which was scribbled a note:

To THE TRAIL BOSSES:
    This man is a good Indian; I know him personally. Treat him well, give him a beef, and you will have no trouble driving through his country.

IKE T. PRYOR.

    The message brought a roar of laughter from Jackman. "Drive out that Pryor steer," was his order. So Pryor's note cost him a steer, a circumstance considered a prime piece of humor over which the trail drivers chuckled for many a day.

    Under the circumstances sacrificing Pryor's steer was not considered a violation of ethics. Ethics were strong in the cattle country, but they sometimes had peculiar angles, a little difficult to understand. The book *Cattle* by William McLeod Raine and Will C. Barnes contains an incident relating to Shanghai Pierce, a famous drover and ranchman. He rode up to an outfit which had just killed a calf with a D brand—which happened to belong to him. Shanghai looked reproachfully at the boss of the outfit, but not wishing to appear inhospitable, merely said mildly, "The day's coming when every outfit's going to have to eat its own beef."

    "Well, now," said the boss heartily. "Mebbe so, mebbe so. In the meantime this here's mighty good meat. Set down, Shang, an' have chuck with us. I've allus found D beef sweeter than any other." And Shanghai Pierce forthwith sat down and ate heartily of the meat which he had furnished without his consent.

    Pierce was an immense man with a fog-horn voice, a great teller of stories in which his favorite butt was himself. As a boy he worked on the ranch of W. B. Grimes, a Texan, who later became a merchant in Kansas City.

It was before the War and Grimes owned a few slaves. Some wild horses were to be broken, and the rancher put one of his Negro boys, who was apt as a rider, on one of the outlaw animals. Shanghai loved to tell the denouement:

"In about four jumps the Negro was eating gravel and was pretty bad shook up. Just then Mrs. Grimes came out. She was red hot and she sure gave old man Grimes a hoolihanning. 'Why didn't you put Shanghai on that horse?' she scolded him. 'That Negro might have been killed and he's worth a thousand dollars.' "

Trail drivers scrambled to start as early in the spring as possible, because the first herds up the trail had the best grass. Trailing could not begin until the new grass came up since the cattle had to graze their way northward. As soon as the prairies began to grow green, the long, serpent-like strings of steers could be seen heading slowly up the trail, and by June or July the ground far on either side of the route traveled was almost pastured off. Another advantage of early trailing was that it enabled the drover to reach the vicinity of the markets early enough to rest his cattle on the pastures for a month or two before selling them, thus gaining flesh and weight.

Always when a trail herd started its journey, it was nervous, "jumpy," and restless the first few days. Longhorns possessed a strong instinct for locality and they realized they were being driven away from their familiar haunts. The attachment of a cow or steer to its home area is well known among cattlemen. Riding over a wide range cowboys are accustomed to meeting the same individual animals over and over again as they revisit certain localities. When driven from places to which they are accustomed, the cattle, if unimpeded, will wander back. In addition to their instinctive desire to turn back, the longhorns were extremely wild and full of spirit so that they were looking for any excuse to bolt. Besides this they were thrown in with strange cattle from other herds, which always contributes to an unsettled psychology among cows—if cows can be said to have psychology.

Anything which might cause a stampede had to be carefully avoided. Occasionally the regular trail crew was helped for the first two or three days by a second crew of cowboys until the home range was safely behind. It was good judgment those first two or three days to ride as far to the side of the herd as possible, while staying just near enough to be seen. No galloping, no shouting. Especially must no man dismount in view of the herd, for this phenomenon would have been regarded by the cattle as so contrary to nature that they would have been filled with an insane urge to *vamose* toward the farthest horizon.

As a safeguard against stampeding a few old and dignified bulls, too ancient and lacking in vigor to be of further value as breeders on the range, were often mixed with the herd. Frequently these veterans did not reach market, falling out or being killed by the trail-side because they were so slow that they held up the progress of the drive. Even if they did arrive at market little money could be realized on them. But they were valuable during the drive, and particularly in the early stages, because they set an example of dignity and decorum for the rest of the cattle. The young and frisky steers resembled so many school boys, aching for a chance to make mischief, particularly eager for the glorious excitement of stampeding. But the bulls, old and heavy, had passed their frisky years. They had no desire to run unless they saw a very good reason for doing so. Moreover they were afraid of nothing on the prairie, except perhaps a man on horseback, and not always of that. If something sent the steers into a panic, therefore, the bulls, instead of joining the hysteria, would hang back to see what was causing the running. If nothing particularly alarming appeared, they would stop running out of sheer indolence, the steers near them would follow the example, and the fashion would gradually extend through the herd so that the cowboys could gain control again. Soon the long bovine procession would be proceeding once more as if it had not been on the frantic verge of stampeding.

To reduce the animal spirits as well as the homesickness of the cattle, a herd nearly always was driven hard the first three or four days—twenty-five or thirty miles a day, whereas the average pace on the drive was twelve or fifteen miles a day. This was to tire the longhorns so that when they were bedded down for the night they would want to rest and would think twice before bolting. It also helped to break the herd to trail driving. The "breaking in" customarily required a week or ten days, but in the end the cattle knew what was expected of them, what the daily schedule was, and became accustomed to the steady movement and the direction.

XI

HOOFS HAMMERING NORTHWARD

A U cowboy rode his tired horse into Kiowa, Kansas, one spring day in 1889 and handed to a telegraph operator at the Santa Fé depot a note:

Will be in Kiowa in ten days with 2,500 sea lions. Have 100 stock cars ready. SNYDER.

The message, transmitted to the general offices of the railroad, created mystification and the suspicion that Colonel Andy Snyder, who had signed the note, was making sport of the Santa Fé. Stock cars for sea lions? It didn't make sense. So a cautious query went back to G. W. Rourke, the Kiowa station agent. Rourke, who should have understood the allusion, was also puzzled by the "sea lions." He found Tom Wilson, another cowboy from the U ranch owned by Major Andrew Drumm and Colonel Snyder down in the Cherokee Strip. Wilson, later owner of the Drag 4 outfit in Arizona, was able to interpret.

"That's what Snyder calls longhorn steers," he explained with some astonishment at the lack of knowledge displayed by the Santa Fé. "They've done so much swimming on the way up from Texas that he calls 'em sea lions."

The surprising thing is that this common slang expression should have puzzled either Rourke or the railroad, because it was in use as early as 1868. Fording or swimming rivers was one of the most difficult, exciting, and dangerous jobs of the men who worked the trail herds north. In the spring the rivers generally were up and it sometimes seemed the journey was nothing but a succession of putting herds across the water. On the Chisholm trail even after the Colorado, Brazos, and Red Rivers were crossed, there remained the Washita, North and South Forks of the Canadian, Cimarron, Salt Fork, Arkansas, and Little Arkansas before the Kansas railheads were reached. No bridges or ferry boats existed. Bank-full the roily, swirling rivers were fearsome, but the cattle had to be put over. Colonel D. H. Snyder's "lead steers"—referred to before—never were duplicated. It required courage and coolness to cross a bunch of longhorns over a river like the Cimarron when it was in flood without heavy losses in cattle or even deaths among the riders. But the cowboys on the trail without hesitation went down to each stream as they approached it, the point riders keeping the van briskly moving, "crowding them" as the saying was, while the swings brought up compactly the rest of the herd and even the drag was kept in good order. Moving rapidly the cattle approached the river. The idea was to carry them into the water by their own momentum. Inducing longhorns to take the first plunge was always a problem and they occasionally refused to enter the water at all. In such cases it sometimes required days to cross a herd over a river.

If the leaders could be crowded in, a cowboy or two plunged their horses into the stream and guided the cattle toward the other bank. Meanwhile the men behind kept the rest of the herd following closely so no open water would intervene between the animals on the bank and those swimming. As long as there were steers close at hand in the water the longhorns on the bank seemed willing to follow the example of those ahead. Out in the stream the herd looked as if it was composed of disembodied heads, floating on the current with starting eyes, spreading horns and flaring nostrils. Their bodies were entirely below the surface as they swam. "Like a thousand rocking chairs floating on the water," as one old trail driver put it.

Always there was danger that unusual waves, or a floating tree, or a whirlpool, or something else might cause the leaders to stop swimming toward the opposite bank and attempt to return to the shore they just had quit. In so doing they would become mixed with the herd swimming behind them and soon there would be in the water an apparently inextricable tangle known as milling. Desperate measures were needed now. If the mill was not broken the cattle would be swept downstream and drowned because their only salvation was to keep swimming straight across while their strength lasted. Into that mass of saber-sharp horns, struggling bodies, and churning hoofs the cowboys swam their horses, and with blows, curses and cowboy yells, sought to start the cattle again in the right direction. In the melee men were sometimes unhorsed. If they were lucky they caught the tail of a steer or a horse and "rode out"; otherwise they had to swim for it, and swimming among frantic, churning hoofs and horns was perilous. Many a man disappeared in such a mill and was never again seen alive.

Yet every herd was crossed. There is no record of a failure, although on occasion a trail boss might hold his cattle on a river bank for a few days when there was prospect of the immediate diminishing of a flood. Horses were swum across; the chuckwagon was floated over with logs tied to its wheels, or emptied and hauled across by lariats, the contents being floated on a raft. A few outfits were known to take along a small boat in a wagon to assist in river crossings. But the practice never became general and soon died out.

Swimming rivers was only one of the tasks of the trail drivers. On the march the herd proceeded in the following general order: The trail boss set the direction and pace. He had to be a man with great cattle sense, a nose for water, knowledge of the country, ability to handle men, and a strong instinct for direction. Few trail outfits had either a watch or a compass among all their men. Under such circumstances it was customary each night to point the tongue of the chuckwagon at the north star and take the direction thus indicated next day.

Two experienced men, called "point riders," formed a point of the leading cattle and kept them headed in the right direction. The herd trailed behind this point, widening out toward the rear. At regular intervals on the flanks came cowboys called "swing" or "sideline" or sometimes "line" riders. At the rear of the herd rode one or more men sometimes called "tailers," but more commonly known as "drag riders." Their duty was to keep the laggard cattle within some reasonable distance of the leaders. Position on the order of march indicated a man's rank. The point riders were the top hands. Theirs was a heavy responsibility, particularly at river crossings and in trail emergencies. Other men graded down as the rear of the herd was approached, the drag being usually the province of tenderfeet or greenhorns. Riding the drag was the most unpleasant work because there a man had to eat the dust from the herd. Some changing about, of course, occurred and the status of the men was by no means inflexible. On hot days, for example, a herd of cattle generated tremendous heat and cowboys on the windward side frequently would exchange places with their sweltering comrades on the heated lee. Similarly drag riders sometimes traded with swing men to breathe clean air for a while.

After the first day or so the cattle evolved gradually a quite regular order of march among themselves. Different individuals fell day after day into accustomed places. The same steers generally showed up each morning in the point. Coasting along the herd a man who knew the cattle could pretty well tell at what section of it he was—front, center, or rear—merely by observing the animals, because, within limits, they would be much in the same places each day.

At a distance to one side of the cattle herd, or sometimes in front or even in the rear, came the horses of the remuda, sixty to one hundred of them, usually herded by a boy, an apprentice as it were, known as the horse wrangler. The chuckwagon was put in motion after the herd was started on its day's march, passed around the cattle and preceded them so that the cook might have camp set up and supper ready when the men arrived at the selected stopping place.

The horses were either herded or hobbled at night. Every cowboy kept one particular horse for night herding. For that exacting service an animal was required with responsiveness, quickness and speed, as well as intelligence, because night stampedes were the terror of the trail. Twice a day a rope corral held the ponies while the riders roped and saddled new mounts. It is really remarkable how cow ponies recognize as a "fence" a series of lariats strung from chuckwagon wheels to slender posts about breast high, to form a circular corral. Sometimes men are pressed into service as "posts" to hold up the lariats. There seems to be a tacit understanding between men and horses that the rope corral which the ponies can easily brush aside is by rule a tight fence, and the little horses play the game according to the rules, today as they did in the early days.

The cowboy's day on the trail started at dawn and lasted through the daylight hours, including a part of the night, and occasionally, in emergencies, the entire night. Two grayed veterans exchanged recollections at the Texas Trail Drivers' convention a few years ago.

"Jim Blocker was the outdrivin'est man with a herd I ever see," commented one.

"Up in the Territory a feller told me the only kick he had driving with the Blocker outfit was that he had to eat two suppers every night," said his friend.

"Two suppers?"

"Yep. One after dark and the second befo' sunup next mawnin'."

Long before dawn the voice of the cook could be heard with his cry, "Chuck-away," or "Come an' git it," or "Arise an' shine." Out of their blankets rolled the cowboys and went to the chuckwagon for breakfast, a simple meal usually, of sour-dough biscuits baked in a Dutch oven, with meat, gravy and black coffee. Vegetables were a rare treat. They were obtained only when trading posts were passed on the trail. Potatoes and onions, the favorites of the range, usually were chosen. Otherwise the diet was monotonously the same.

After breakfast the men went to the rope corral into which the remuda had been driven; there caught up and saddled their forenoon horses. The last night herders were now relieved and ate their breakfast while the cattle were put in motion. For the first two or three hours the longhorns were permitted to graze. Even as they grazed, however, they were so guided that they moved in the desired direction. After grazing two or three miles they were started on the trail and driven briskly for three or four miles more by noon. Then, if water was available, they were permitted to drink their fill and either graze or lie down to rest through the warmer hours. The cowboys took advantage of this period to change horses and perhaps even to catch naps in the shadow of the chuckwagon, if they were not on active duty. Toward evening the herd once more was set in motion and driven seven or eight miles, the average full day's trip being about fifteen miles—after which they were allowed to graze until night and then put on the bedding ground.

It is interesting that settlers often requested that herds be bedded down on their property. This was because a large herd of cattle would leave a quantity of "cow chips" which, when dried, was the traditional fuel of the plains. One Kansas nester estimated that a single herd, in one night, left him five hundred pounds of "chips"—almost enough for his winter's fuel supply.

Cattle must be eternally watched. Throughout the night the cowboys, in shifts divided up equally so that each man watched about two hours, rode continuously around the herd. Two usually were on guard at the same time, traveling in opposite directions. As they rode slowly they hummed or sang. It was necessary to keep up some kind of sound. Cattle enjoy being sung to, no matter how discordant the music. Besides, a continuous sound of some kind reassures the animals, and keeps them apprised of the position of the rider, so that there is no danger of startling a sleeping animal into leaping to its feet with a bound and a snort, thus perhaps setting the whole herd off in a blind smother of galloping hoofs and clattering horns. At the end of two hours the second watch relieved the first, and so on throughout the night. This was everyday routine, and the trail drivers thought nothing of their extra duty after fourteen hours a day in the saddle.

No tents were provided for sleepers. Bedrolls were spread outside, under the stars—when there were stars. If it rained, and it often did, sleeping was wet and uncomfortable and there was little of it. Stormy weather was the danger period for stampedes. James Cook, author of *Fifty Years on the Old Frontier*, tells of one drive when the entire prairie was so swampy that the men could not sleep in the ordinary manner. When three of them could slip away at the same time, they would dismount, each man holding his horse by the bridle rein. Then they would lie down in the form of a triangle, each man using his neighbor's ankles for a pillow to keep his head out of mud and water.

In long rainy periods men grew worn almost to the limit of endurance. Occasionally cowpunchers went to sleep in the saddle and there was danger of falling and injury. What was known as a "rouser" was the placing on the inside of the eyelids of a little tobacco juice. It caused smarting so painful that it banished drowsiness, but was extremely irritating to the eyes. The heroic treatment indicates how serious was the situation that demanded it.

Need there was even for the cowpunchers asleep in their blankets to cultivate an ability to awake to instant alertness. Stampedes were the great danger of the trail; especially night stampedes. Storms, particularly thunder and hail storms, were dreaded because they often caused the cattle to bolt. So wild were the longhorns that anything, almost, would set them off. A setting hen which flew cackling from her nest once sent several thousand steers on a desperate and costly stampede. Scratching a match to light a cigarette near a herd at night, or a sudden sneeze, or the crackle of a branch stepped on by a pony, or even a vagrant tumbleweed rolling along in the breeze, might launch a disastrous flight.

The insidious thing about stampeding was that it bred the tendency to stampede again. Herds became "spoiled." There is a record of one herd on the Chisholm trail which stampeded eighteen times in a single night. Sometimes the runs were short; at other times stampedes became all-night affairs. A bunch of little brush cattle from southern Texas has been known to run in a line for forty miles. They were always more difficult to handle than the bigger cattle farther north. Always some animals were lost, killed or injured, and time was invariably lost seeking strays after a bad stampede. The panic-stricken flights therefore were economically serious.

Frequently a herd contained a dozen or so old, wild steers which developed the stampeding habit. These were continually on the lookout for some excuse to run, and if no legitimate excuse presented itself, they created one. At any moment they were ready to go unreasoningly into flight, carrying the rest of the herd with them—apparently for the sheer exhilaration of the mischief they were doing. Usually the chronic stampeders grew to know one another, and the cowboys would notice them hanging around together as if plotting fresh outbreaks. But the wrong-headedness of these animals sometimes was their downfall. Just so much of their particular brand of insanity would be stood by the trail boss; then he would cut those steers out, give them to a band of Indians, sell them if he had a chance, or, lacking other expedients, drive them off and shoot them. Even the last extremity was economy in the end, because a herd of steers which was always stampeding lost weight rapidly, and the removal of the disturbing influences paid for itself in added tallow on the cattle that were left.

Nothing in the West exceeded the portentous circumstances building up to a typical stampede. The herd of cattle would be bedded down, but the cowboys circling about and chanting cattle lullabies would notice that they did not lie fully relaxed. Some crouched with their legs drawn up, as if ready for a sudden spring. Pitch black was the night except for lightning playing on the horizon, and the distant mutter of thunder gave an added ominousness to the scene. Occasionally a steer would rise, stand quivering with tenseness, and sniff the air. All about the bedding ground, the cowboys knew, ran crooked gulches and gully washes, slashing across the country which also was studded with prairie dog villages, the bane of a running horse, night or day.

The air became filled with electricity. Sometimes it was so charged that an equivalent to what is known on the ocean as St. Elmo's fire appeared. At sea this is observed as luminous bulbs of light at the ends of spars on a wet, storm-threatened night. Old cattlemen will tell you of seeing the same thing in a herd of cattle, small levin-lights at the tips of the horns on every beast's head giving a weird effect.

Suddenly the whole dark world blazed in the white light of a blinding flash, accompanied by a terrific crash of thunder which echoed and re-echoed, dying down in earth-shaking reverberations. But the final rumble was not heard. It was drowned in a new, tumultuous thunder of thousands of hoofs all drumming at once, of bawling cattle, of clattering hocks and clashing horns as the herd leaped in a single, upsurging motion to its feet and bounded off into the darkness of the now falling rain.

Lucky was it for the camp outfit if the stampede did not start in its direction. The men, watching with foreboding the black night, lay with their horses saddled, bridled and close at hand. No sleeping in that bivouac. As the thunder-clap was heard and the cattle roared into flight, every rider, without waiting for orders, vaulted into his saddle and rode at his horse's hardest after the herd.

No use trying to head off the fear-maddened beasts. They would run down and trample underfoot anything which came into the way of their blind, terrible rush, although in daylight, and unfrightened, cattle will leap over an object in their path. The only hope in a stampede was to follow the herd at the top speed of your horse, over prairie dog holes where a step into a burrow meant a broken leg for the horse and a shattering fall for the man; skirting yawning gulches or plunging blindly down into them when they could be crossed; trusting utterly to the sureness of the cow pony, for in the blackness the hand could not be seen before the face, and the cattle, once in motion, had lost their levin-lights and could be followed only by the tumultuous roll of their hoofs.

Gradually a stampeding herd always strung out, as the fastest animals forged ahead, until the chance came for the cowboys riding abreast. The most successful way to break a stampede was to swing the leaders into a wide curve so that eventually they were running in a circle, winding the circle closer and closer, until at last the herd was milling, entangled in itself. Very often, however, this was impossible and the animals had to be allowed to run as long as their endurance lasted.

After a stampede the riders always made a quick check

to see if all hands were present. Frequently it was necessary to ride back with the dawn over the stampede trail,
to pick up what was left of a companion who had been
battered by hundreds of hoofs. The faces were wan in
the morning, and the spade in the chuckwagon was brought
to its appointed use. Then, with no gesture even at a funeral service, the body would be lowered into the grave
and the drive would continue once again toward the market at the dim end of this bloody trail.

XII

JOSEPH McCOY'S GREAT IDEA

One of the men who made the ill-fated drive of 1866 was W. W. Sugg, an Illinois cattle dealer, who went to Texas to buy steers cheap and sell them high in the North. The Jayhawkers neglected to differentiate between a trail driver from Illinois and one from Texas. All were equally fair game, and by the time those freebooters finished with Sugg, he had little left but a knowledge, made vivid by adversity, of conditions, trails, and men in the cattle country.

That knowledge, passed on to another man, produced a new historical epoch. Sugg knew a young cattle dealer in Springfield, Illinois—Joseph G. McCoy, a visionary, dreamy individual, inclined to be verbose, but of a sympathetic nature withal. After his harrowing experience at Baxter Springs, it was natural for Sugg to recount his doleful story to McCoy. He described the immense numbers of cattle in Texas, practically worthless because they lacked a market, and told of the impossibility of bringing them to the North because of granger and Jayhawker opposition. As he talked, McCoy listened with mounting interest. Sugg's words tremendously stimulated his imagination. At first he confided his inspiration to nobody. Later he described his own mental processes in his *Historic Sketches of the Cattle Trade*, which even today is a foundation reference work on the trail drive period:

This young man [McCoy refers to himself in the Caesarian third person] conceived the idea of opening up an outlet for Texan cattle. Being impressed with a knowledge of the number of cattle in Texas and the difficulties of getting them to market by the routes and means then in use, and realizing the great disparity of Texas values and Northern prices of cattle, he set himself to thinking and studying to hit upon some plan whereby these great extremes could be equalized. The plan (finally decided upon) was to establish at some accessible point a depot or market to which a Texan drover could bring his stock unmolested, and there, failing to find a buyer, he could go upon the public highways to any market in the country he wished. In short, it was to establish a market whereat the Southern drover and the Northern buyer would meet upon an equal footing, and both be undisturbed by mobs or swindling thieves.

Stated in those terms the plan sounds simple. Actually it was beset by tremendous difficulties. Yet it was one of the great single ideas of this nation's history. Coming at the time it did, it rescued Texas from bankruptcy, launched a tremendous migration which within two decades populated the West, and gave an historic savor to the greatest section, territorially speaking, in the United States. McCoy's dream was of the stuff of which empires are created.

Pursuing his idea, McCoy visited Kansas in the spring of 1867, traveling out on the Kansas Pacific railroad which was just pushing across the plains. His cattle depot must be situated west of the settlements so the trail drivers would not again be blockaded in the northern course. He must establish shipping facilities, provide transportation for cattle brought to his depot, interest trail drivers in his market, and induce Eastern buyers to come to his rendezvous. The task would have disheartened anyone with less volatile spirits and high imagination than McCoy.

At Salina, Kansas, he found the railhead of the Kansas Pacific, but when he attempted to secure land there for his stockyards he found the inhabitants not only cold, but hostile. The valley had been populated by a colony of Scotch Presbyterians, and all they wanted was to till the land and fear God. They seemed to feel that both these programs would be interfered with if the unregenerate horsemen from the South came into the country. Later McCoy was refused locations at both Solomon and Junction City. Finally he decided upon Abilene, a little, unprepossessing Kansas hamlet of a dozen ramshackle log huts, with a single street so lacking in traffic that a busy and populous prairie dog town existed in the middle of it.

For all its unimpressiveness, Abilene had one or two important advantages for a cattle market. On the railroad, it was sufficiently west of settled country to allow cattle to be trailed without interference. It lay in the Smoky Hill valley, a smiling depression in the plains a mile or more wide, with gently sloping sides, trees in the bottom, and covered with grass so rich that it impelled Tim Hersey, the chief land owner of Abilene, to remark to Theodore C. Henry, later first mayor of the town:

"See them prairies all around here? I'm losing a million dollars a year on 'em."

"A million dollars?" inquired Henry. "How's that?"

"By not havin' a million head of sheep grazin' on 'em."

Other advantages of Abilene were that it was not far from Fort Riley, which provided a market for beef to ration soldiers and at the same time afforded protection from the Indians if need arose.

Before definitely settling on the site McCoy went east to talk business with the railroads. He found them lacking his own vision and enterprise. The Kansas Pacific was so dubious that its president said his company would risk not one cent on the project, and agreed only as a concession that if McCoy carried out the plan the railroad would furnish a switch. The president of the Missouri Pacific would not even listen to the proposition and thereby changed history, because the Hannibal & St. Joseph railroad quickly entered into a contract, with satisfactory freight rates, to ship cattle through Quincy to Chicago, laying the foundation for that city's great packing industry, while St. Louis missed becoming a primary cattle market. One advantage McCoy wrested from the Kansas Pacific: he obtained a contract whereby he was to receive one-eighth of the gross amount of freight on all cattle shipped east over the line. Even this contract eventually was abrogated and McCoy as a result was bankrupted. But that is personal history and has no place in the larger events of the cattle industry.

Having formed a company and secured financial backing, McCoy arrived in Abilene in June. By July 1 he had commenced the construction of a shipping yard and offices.

It was no ordinary cattle pen he designed, but a yard built to hold the surges of three thousand wild cattle, with posts of railroad ties braced by other ties, and with planking stout enough to resist the rushes of the heaviest bull. In addition McCoy erected a three-story hotel, the famous Drover's Cottage, and an office, with a large Fairbanks scales, of which he was inordinately proud. Within the remarkably short period of two months the work was finished and McCoy was ready to start business.

While all this was going on the young promoter was not idle in other fields. He employed Sugg—the man whose mournful story had been the first inspiration of this project —and sent him south to meet the trail herds and turn them

toward Abilene. The following spring he sent Colonel
Hitt, Charles Gross and possibly another man on the same
errand.

Many herds were working northward at the time, although not so many as in 1866, owing to the disappointments of that ill-starred year. Several were already crossing
the Indian Territory, hoping to run the gauntlet past Baxter Springs. It was Sugg's job to turn them west toward
Abilene. This he succeeded in doing in numbers sufficient
to bring 35,000 cattle for shipment out of McCoy's yards
in 1867. But the first herd into Abilene came there by sheer
luck, and by sheer accident the great Chisholm trail, most
famous of all the cattle routes, was established.

Colonel O. W. Wheeler was an adventurer of the authentic early American pattern. He had hunted gold in
California and later entered the livestock business there. In
1861 he guided across the plains the greatest emigrant train
known to history—a caravan containing between six hundred and eight hundred persons, with several hundred
wagons, and more than one thousand head of loose stock
in addition to the work animals. The Apaches, as usual,
were on the war path, but Wheeler's army of people, animals, and vehicles was so immense and well-conducted that
the savages were overawed and it is said that not a single
head of stock or a single human life was lost in this mammoth hegira.

A drouth in California in 1866, coupled with a dearth of
cattle, created a demand for stock animals there. Wheeler
knew there were longhorns in immense numbers and very
cheap in Texas. With his partners, Wilson and Hicks, he
crossed the plains and purchased twenty-four hundred cattle to trail back to the Pacific coast. Next he employed
fifty-four cowboys, all of whom had to qualify not only
as herders but as Indian fighters. These were equipped not
only with the ever-present six-shooters, but with Henry

repeating rifles, the predecessors on the plains of the Winchester.

But that herd of Wheeler's never reached California. It left San Antonio early in the summer and struck north for the California trail along the Platte by way of South Pass. Wheeler, apparently, knew nothing of Abilene and its pretensions when he started, but as fate ordained his lead steers pointed their noses straight at McCoy's ambitious new enterprise. As the drive approached the Kansas line reports came that the terrible Asiatic cholera was sweeping the West. All the way up and down the Missouri River and across the great trade routes, men were dying by scores from the sudden, mysterious disease. There came, moreover, new disquieting rumors of hostile activities by the Indians. The Cheyennes were playing tag with General Hancock's clumsy expedition in Western Kansas and tearing up the tracks of the Union Pacific in Nebraska. By the time Wheeler reached central Kansas his partners were thoroughly apprehensive and declined to go farther west. Angered at the defection the colonel held the herd near Abilene, too stubborn at first to sell it.

Thus he lost the honor of driving the first cattle to McCoy's yards. A second herd, belonging to a man named Thompson, had crossed the Red River and followed Wheeler's trail. Thompson sold out in the Indian Territory to a firm of speculators, Smith, McCord & Chanler, who took it on north, passed Wheeler while he was quarreling with his partners, and launched Abilene's historic business boom. Later Wheeler brought in his herd and other drovers came, but not as many as McCoy had hoped.

The truth is that 1867 was a bad year for the cattle market. Heavy rains caused the grass on the trails to grow coarse and washy, with little nutriment. Rivers were flooded and stampedes were unusually frequent. When the longhorns reached Abilene they were lean and worn.

Nor were these the only complications. In the East the corn crop was almost a failure and few farmers wished to buy stock as feeders, so the thin, raw-boned Texas cattle had to go directly to the butchers' markets. An unreasonable prejudice, also, had developed against Texas cattle in the East. The statement was freely circulated and believed that longhorn meat was "about as palatable as that of prairie wolves." It was untrue, of course, but it had a reasonable sound and people always have been ready to accept reasonable-sounding phrases at face value. Had the Texas cattle gone to the markets that year of 1867 in plump condition, the "prairie wolf" canard would have been dissipated. Unfortunately the cattle were thin and the prejudice festered. No profit was made on the first shipment from Abilene to Chicago. The second went to Albany, New York, a trainload of nine hundred head, and sold at $300 less than freight charges. No more cattle went to New York state that year and the situation looked bitterly discouraging.

Yet with unflagging zeal, McCoy already was making plans for 1868. Representatives were sent to Texas to talk with prospective drovers and the entire cattle country was circularized. Advertisements were placed in Eastern newspapers to attract buyers to Abilene. The railroads, at last awake to their opportunity, assisted McCoy by publicizing their rates and routes.

As if his other problems were not enough, McCoy faced a delicate matter of diplomacy at Abilene. Still in the minds of the grangers was the bugbear of Spanish fever and with the new bustle and beginnings of prosperity at Abilene many families had moved in and taken farms in the vicinity. Some of these for various reasons—which might not in certain cases have borne analysis—organized an "association" to prevent Texas cattlemen from driving their herds to McCoy's stockyards on the ground that the longhorns damaged crops, spread the dreaded disease, and were dangerous. Plans were even made to stampede the cattle herds as they came. In his reminiscences McCoy tells how he dealt with these obstructionists. Calling a meeting of the grangers, he made a speech, pointing out the advantages of the large cattle trade to the community, and suggesting that they all could make direct profits by selling produce to the drovers, and by investing in young, cheap cattle. His account continues:

Whilst this little talk was being made, nearly every drover present, by previous arrangement, went to bartering with the Kansans for butter, eggs, potatoes, onions, oats, corn, and such other produce as they might be able to use at camp, and always paying from one-fourth to double the price asked by the settlers. At the conclusion of the meeting, the Captain [leader of the grangers] said he had got a "sight" of the cattle trade that was new and convincing to him. "And, gentlemen," said he, "if I can make any money out of this cattle trade, I am not afraid of Spanish fever; but if I can't make any money out of this cattle trade, then I am damned afraid of Spanish fever." The entire hostile organization dissolved without any further trouble and before a single steer was stampeded.

Yet the opposition remained in a quiescent form. It was a thing more fundamental than McCoy realized; the old enmity of the husbandman for the herdsman which has existed since Cain and Abel. It was destined to continue to the end in cattle land.

Jesse Chisholm was a half-breed Cherokee who capitalized on the canniness of his Scottish father and the racial contacts of his Indian mother to make a success of trading with the plains tribes. For him the Chisholm trail was named, but it should by rights have been named instead the Emory trail. It is true that the first trail herds coming north in 1867 followed a route over which Chisholm had driven his wagons and some livestock south from the present site of Wichita, Kansas, to trade with the Indians near Fort Cobb in 1865. That route touched the best fords and water holes across the central part of the Indian Territory. Chisholm himself, however, followed an earlier track

made by a federal military force commanded by Lieutenant Colonel William H. Emory in the first days of the War
between the States. In 1861 the Confederate troops were
converging from Texas and Arkansas northward in the
Indian Territory, menacing Union forces garrisoning Fort
Washita, Fort Arbuckle and Fort Cobb. Emory was in
command of the military area. With great decision and
enterprise he mobilized his scattered forces for a march
across the unmapped plains to the north, hoping to reach
the federal lines. All other routes were blocked.

A famous old Delaware scout named Black Beaver
guided him. This Indian had previously blazed the trail for
Audubon, Kearny and Marcy in their western expeditions.
With Black Beaver leading, Emory marched his seven
hundred and fifty soldiers and one hundred and fifty non-
combatants north over the best fords of the Canadian,
Cimarron, Chikaskia, Ninnescah and other rivers to the
Arkansas, where, at the present site of Wichita, he dispatched his first message of months to military headquarters announcing the completion of his march. Chisholm
back-tracked Emory's trail after the War, when he drove
his wagon train south, and the cattlemen following Chisholm's wheel tracks established the cattle route. History
has a way of giving its honors to the wrong men.

Whatever the merits of its name, the Chisholm trail was
the most famous of the cattle roads and ranks with the
Santa Fé and Oregon trails in the history of the West.
Years after it ceased to be used traces of it could still be
seen stretching across unplowed areas in the Indian Territory, sometimes two hundred to four hundred yards wide,
a depression beaten by the hoofs of hundreds of thousands
of animals, and further eroded by wind and rain in the
years when it was a great thoroughfare of empire. Bleach-
ing skulls and skeletons of brutes which had fallen by the
wayside marked its course. Low mounds indicated graves
at frequent intervals, and there were broken-down wagon
frames and the traces of circle-like bedding grounds on

either side which could be distinguished. Nobody knows exactly how many cattle were driven up that route, but the number is estimated at from three to five million. Other routes in the West saw great migrations of horned beasts, saw cowboys by the thousand on their tough little bronchos, saw courage and tragedy and hardship and crime, but the Chisholm trail, running from Red River Station to Abilene, was the greatest in history because it first caught the imagination of the nation and gave substance to the tradition of the trail.

Having advertised for drovers in Texas and buyers in the East, McCoy was rewarded by seeing Abilene do an immense business in 1868. More than a thousand carloads of cattle were shipped in June. As swiftly as it had risen the Eastern prejudice against Texas beef abated, and the market boomed. Other Kansas towns, observing the spectacular success at Abilene, began to compete with it. Yards were constructed in one city east and three west of Abilene, and men were sent from these rival towns to the crossing of the Arkansas River to deflect herds their way. But McCoy was equal to the challenge. Once more he sent Sugg down to the drovers, and the message this time was: "Abilene not only has yards and accommodations—*it has the buyers.*"

T. E. Hersey, who was an engineer, straightened out the trail from the Kansas line north, marking it with mounds of dirt. So desperate grew Abilene's rivals that one town—unnamed in McCoy's account—even tried to bribe a drover with an offer of $600 in cash to take his cattle there. In spite of this herds continued to pour to Abilene in unchecked thousands; McCoy's yards had more than they could handle; merchants were busy and everybody was prosperous.

And then there was a disastrous occurrence. A Chicago firm tried the experiment of shipping thousands of cattle by steamboat from Shreveport down the Red River and up the Mississippi to Cairo, Illinois. The cattle were lousy with ticks and debilitated from their long water journey, and they quickly spread disease among the native cattle where they were distributed as feeders. At Tolano, Illinois, the scourge carried off nearly every animal in the countryside. Other communities were affected and throughout Southern Illinois livestock owners became panic-stricken and began shipping their cattle out of the state to avoid losing them. Once more Texas cattle were under the interdict. The market went into convulsions and longhorns were banned. Many herds, reaching Abilene, found there was no sale for them at any price.

Once again McCoy rose to the emergency. This time he conceived the idea of sending a shipment of buffalo east to call attention to Abilene and popularize the plains. Bison herds grazed directly west of the town and, having prepared a stock car by strengthening it with bolted planks on the inside, McCoy took a party to Fossil Creek, sixteen miles from Abilene, where the car was run out on a siding. The promoter had his choice of some of the finest ropers in the world. A group of California *vaqueros*, celebrated with the lariat, happened to be at Abilene, and of these he selected three, together with four Texas cowboys. What followed when the little band of noose experts started for the buffalo herd must have been exciting.

The method was as follows: Two men would rope a buffalo about the neck and gradually work him toward the car, keeping him from charging in either direction by their taut ropes. When he was maneuvered near the car a third lariat whipped about his heels, he was thrown and hog-tied, and with a block and tackle was hauled up an incline of planks into the car where his head was fastened securely before his leg bonds were released. The miracle was that the cowboys and *vaqueros* actually captured ten full-grown buffalo bulls. Of these, four died from heat and excitement, three lay down and refused to rise so that they could not be brought near enough to use the block and tackle, and only three were brought to Abilene.

With a flair for showmanship which would have made a fortune for him in the circus business, McCoy caused large canvases to be painted with gaudy signs proclaiming Abilene the greatest shipping point for Texas cattle, and had them hung on the sides of the car. Then the buffalo were sent to Chicago by way of St. Louis. They created tremendous excitement and achieved exactly what was desired. Abilene once more became the focus of interest for the nation's cattle buyers.

Already, however, the end was almost at hand. Prosperity brought a growing population to Abilene and the newcomers began to feel a smug disapproval of the uncouth Texas cowboys and of the gambling, liquor and red light district which the wild youths from the cattle country so joyfully patronized. Civic pride burgeoned, and in February, 1872, a group of citizens, unbelievably blind to their own or their town's interests, drew up a circular and sent it down to Texas, notifying the cattlemen to drive to another point.

Too late some of the clearer thinkers of Abilene realized what had been done. Another circular was hastily issued and signed by some of the very men who had signed the first manifesto, imploring the Texas drovers to ignore the first circular and return to Abilene with their herds. But it was too late. As quickly as a stone is dropped into a well Abilene was dropped by the cattlemen. Thereafter it retired to a not unmerited obscurity.

XIII

OCEANS OF GRASS

In the schools of the '60's every urchin, yearning to leave his hard bench and seek the congenial freedom of the out-of-doors, knew there was a Great American Desert. He knew it because it was there in print before him—on the pages of his geography, with boundaries well defined and quite explicit: it was bounded on the north by the upper Missouri, on the east by the middle Missouri, on the south by Texas and on the west by the Rocky Mountains. Within that immense area, the schoolboy was informed, no human civilization could subsist.

It was known, to be sure, that grass covered most of this territory and some conclusions might have been drawn from the fact that buffalo herds numbering millions of beasts and antelope, deer and other wild ruminants subsisted and grew fat upon that grass—the wonderful, short, curly buffalo grass and the bunchy grama grass, both of which had the quality of curing in the summer into natural hay on the root and retaining peerless nutritive values throughout the winter. Many stockmen hold today that the Western grasses are better for stock after curing and assuming the yellow-gray appearance of being dead than when they are fresh and green in the spring.

But the Great American Desert legend, disseminated by the school geographies, lasted for a long time, and the realization that the high plains of the north were a superb natural cattle range, given normal weather in winter, came slowly. There is a story, frequently repeated, with the pleasing semblance of romance, concerning this discovery of the plains. In December, 1864, so it goes, a freighter with a load of supplies drawn by oxen was caught in a bliz-

zard on the Laramie plains of Wyoming, while on the road
to Camp Douglas in Utah Territory. There was no way
for him to care for his oxen, so he went into winter quarters and turned the animals loose, expecting them to freeze
or starve to death. To his amazement, when spring came
he found the oxen not only alive, but in a fine condition
of flesh due to the excellent grass on which they had pastured. Thus, presto, the northern plains were discovered
as cattle range.

Unfortunately for this story the value of the northern
range for winter grazing was known long before 1864.
Freighting had been in progress for decades by that time,
and oxen were common, if not favorite, work stock. One
Seth E. Ward, a settler of Fort Laramie, is credited with
being the first to winter a cattle herd in southern Wyoming. During the cold months of 1853 he left his cattle in
the valleys of the Chugwater and Laramie and found that
they survived in good condition. The following year, 1854,
Alexander Majors, the great genius of Western freighting,
delivered 100,000 pounds of freight at Fort Laramie in
November. It was too late to return to the Missouri and,
having learned of Ward's experience, Majors decided to
winter his three hundred work oxen. He had them herded,
however. Although the oxen went on the range sore-
footed, yoke-galled and emaciated, they came out the following spring in the finest working condition. For ten
years thereafter Majors continued to winter his cattle in
those ranges with excellent results.

A similar experience was that of Captain W. F. Rayn-
olds while exploring the Yellowstone River in 1859. Arriving in the valley of Deer Creek in November, he went
into winter camp and turned loose to shift for themselves
his seventy horses and mules, all of which were very thin
and worked down. In the spring he was surprised to find
them in condition as good as if grain-fed and stable-housed
all winter. In his report to the United States Senate he
wrote that this incident was "the most forcible commen-

tary I can make upon the quality of the grass and the character of the winter."

Year-around pasture on the high plains, both north and south, existed by virtue of the climate and the vegetation which had developed in it. Enough rain generally fell in the spring, added to the melted winter snow, to give the grass a quick early-season growth. As summer waxed, however, the precipitation ceased and eventually a drouth exhausted all moisture. Meantime the atmosphere was so dry that the grass cured. In damper climates grasses continue to grow until frost kills them, and they then become woody, lose their nutritive elements and usually are unfit for grazing.

When Nelson Story took his trail herd up the Powder River to Montana in 1866, the beginnings of a range industry were already springing up there. Mining camps needed beef and Indian reservations and military posts needed beef. Railroads, those antennae of civilization, were feeling their way westward and the construction camps needed beef. To supply this many-sided need, little cattle ranches began to appear in the mountain valleys of western Montana. Most of the cattle which stocked them were brought in from Utah or Oregon. For the present the full scope of trail driving could not begin from Texas because the Sioux still asserted their dominance of the Wyoming plains, the Black Hills and the western Dakotas. The Sioux formed a barrier to the cattlemen—a barrier studded with lances and arrow points and lethal intent.

But the cattle range had need of spreading and already it was looking for a place to spread. To Abilene, 1871 was the high tide of trail driving. An estimated six hundred thousand longhorns crossed the Red River that year, most of them bound for the railroad. Ellsworth, Newton and other towns shared this shipping business with Abilene. But that year the cattle market at Chicago and elsewhere collapsed for economic reasons, while the railroads entered

into a combination to raise freight rates from Chicago east.
Therefore as the season grew older, the Texas drovers
found buyers steadily fewer. Cattle brought north that
spring, moreover, had not done well. It was a wet, stormy
season, and the grass grew rank, spongy and woody along
the trails. Steers arrived in Kansas scrawny and weighing
light so that they brought low prices. Even cattle held on

the prairie failed to fatten and by fall all shipping had
ceased and the markets were glutted.

Meantime the slow herds from Texas continued to arrive. When they reached Kansas there was no place for
them to go. The cowboys had accomplished the tremendous task of bringing them north, swimming rivers, riding
out stampedes, enduring storms, braving Indian perils—and
here was no reward at the end of the trail. It was heartbreaking, especially since some of the drovers were unable
to pay off their men. Yet with the touching loyalty which
so often characterizes the men of cattle land, many of the
riders from the South stayed on to care for the herds of
their bankrupt outfits. For lack of room about the market
towns the herds spread westward exactly as water spreads
over flats when it is dammed at the outlet. The *Saline
County Journal*, published at Salina, Kansas, said on July
20, 1871:

The entire country, east, west and south of Salina down to the Arkansas River and Wichita is now filled with Texas cattle. There are not only cattle "on a thousand hills" but a thousand cattle on one hill and every hill. The bottoms are overflowing with them and the water courses with this great article of traffic. Perhaps not less than 200,000 of them are in the state, 60,000 of which are within a day's ride of Salina, and the cry is "still they come!"

And still they came, piling herd on herd, and for sheer existence pushing farther and farther west into the buffalo grass country. McCoy estimated not less than three hundred thousand head were put on winter pasture in Western Kansas, most of them held there by their owners, the Texas drovers, and herded by Texas cowboys.

And destruction smote them. Out of the north swept a white, blinding smother—a blizzard, the first the Texans had met. At first rain fell, freezing as it came down until the earth was sheathed in ice two or three inches thick and the short buffalo grass was completely covered. Then came a furious gale of piercing wind, sending the thermometer plummeting and bringing with it a thick mist of driving snow and icy particles against which it was impossible to travel; it was necessary to turn the face down wind in order to breathe.

It was murder to the Texas cowboys in their thin Southern garments and their leathern chaps and shirts. Many of them died in the howling storm. It was murder also for the Texas cow ponies, used to a warmer climate, who turned their rumps, sulked and groaned, and finally in many cases went down to freeze to death. But particularly it was murder to the Texas cattle. A cow is especially helpless when the grass is covered. Unlike the horse she cannot eat snow for water, nor does her instinct impel her to paw the snow, as does the horse, to reach the grass. Sometimes cattle attach themselves to a band of horses, following after and keeping life in their gaunt bodies by eating the few spears of grass left by the horses when the latter have pawed up

the snow and filled themselves. In the winter of 1871 the
sheath over the grass was even worse than usual, because
the first coating on the ground was solid ice. Only one
thing there was for the cattle to do. They turned their
meek, protesting tails to the blizzard and began slowly to
drift with the wind.

Mile after mile that inexorable drift continued. After a
time animals began to drop out, to lie down, and whenever
they did so they died. For three long days and nights the
storm whooped on; when it ceased and the drovers could
make an inventory, many of them were ruined utterly.
One typical outfit started the winter with thirty-nine hundred head of cattle on the Republican River and in the
spring could find only one hundred and ten of them.
When the thaw came some men made a little money by
skinning the dead cattle and shipping their hides. Small
railroad stations standing solitary on the prairie had stacked
around them, waiting to go to market, ricks of twenty-
five, thirty, even fifty thousand hides. A quarter of a million cattle, hundreds of cow ponies, and nobody knows
how many cowboys died in that great blizzard of 1871. It
was the first wholesale die-up in the northern cattle range.
By no means was it the last.

Meantime the herds trampled their way Southwest and
then west from Texas.

Oliver Loving and Charles Goodnight stole a march on
the rest of the cattle country when they took two trail
herds over the Horsehead route in 1866. But when 1867
opened the partners understood that their monopoly of the
Pecos valley market was ended. Several outfits were making plans to push cattle across the ninety-six mile water-
less trail, including those belonging to Captain Jack Cureton and Jim Burleson. The two daring associates, therefore, were at work early putting together a bunch of cattle, for they had a new plan. They were aiming not only
at the military post market in New Mexico but the stocker

market farther north in Colorado. As soon as the herd could be gathered the drive began toward the headwaters of the Middle Concho, jumping off place for the Horsehead route. Goodnight divided their cattle into two bunches. These were followed by the Cureton herd and preceded by that of Burleson, and another outfit belonging to Ike Cox and John Gamel, of Mason, Texas. We may suppose there was a race between these and other trail crews to be first to the markets.

Early in the trip unmistakable sign on the trail showed that the Indians were out. They had learned of the partners' activities the year before and were lurking in the Horsehead country to stop further ventures of the kind. How bad the Comanches were was not understood, however, until later. They attacked the Goodnight-Loving outfit twice, on one occasion wounding Long John Loving with an arrow which stuck in his skull. Goodnight pulled the arrow point out of the tall youth's head with a horseshoe pincers and sent Long John—who was no relation to Oliver Loving—home to his mother to nurse.

Jack Cureton's herd, which was made up on Jim Ned Creek, a tributary of the Colorado, arrived at where Fort Concho was being constructed and found that shortly before a German had been killed and scalped on the trail and a group of gold miners led by a Captain Snively had fought a brisk skirmish with the same band. After the Horsehead route was passed, the same outfit learned that again they had just missed a war party which killed a man and wounded a woman at Horsehead Crossing trading post, and stampeded the Cox and Gamel herd before mentioned. The Curetons missed the Indians both times and reached Fort Stanton safely.

Loving and Goodnight did not fare so well. Before they left the Middle Concho they knew they were in for trouble. In one respect that was a remarkable drive for the Horsehead route. It was recalled later by one of Goodnight's drivers that "it never failed to rain on us every day until we reached Horsehead crossing." The "long dry march" became a march of mud and for once the cattle did not suffer from thirst. In one way this was not so fortunate. The weather kept the cattle nervous and ready to stampede—which they did the very night the crossing was reached.

Next morning every man in the outfit was out trying to round up the scattered stock. At the conclusion of three days of hard riding a check-up showed three hundred head still missing; and now the trail of the lost cattle was found —with unmistakable signs that Indians had run off that part of the herd. Almost simultaneously word came that the Comanches had attacked and stampeded part of the second herd owned by the partners, escaping with a thousand head.

There had been so much delay that a council of war was held and the decision reached that Loving should ride on to Fort Sumner where the beef contracts were scheduled to be awarded in June. Loving was accompanied by "One-Armed Bill" Wilson, while Goodnight followed more slowly with both cattle herds. It was a dangerous journey the two emissaries had before them. Right through some of the worst Indian country it cut. Over and over Goodnight warned them to travel only by night and hide by day.

Loving, however, was impatient and hated to ride at night. His partner later said of Oliver Loving, "He was a man of religious instincts and one of the coolest and bravest men I have ever known, but devoid of caution." That is an understatement. Loving was reckless to the point of foolhardiness. He had the Westerner's contempt for the Indian which continued until Red Cloud and Sitting Bull wiped it out and Chief Joseph's Nez Percés riveted the new conviction in the long, brilliant campaign from the Walla Walla valley to the Bear Paw Mountains.

Although he must have known that a man had been killed on the trail and Captain Snively's miners had been forced to fight for their lives, it was not long before Loving persuaded his companion that they would make better time by day riding, and also that the Indian menace in the two hundred miles they were to cover, had been over-rated by Goodnight. For two nights they rode, resting the intervening day, without a sign of a Comanche. The next day they traveled by sunlight.

That very afternoon, shortly after leaving the Black River, Goodnight's warnings were justified. A big band of Indians suddenly appeared, cutting the two men off from the direction they wished to go. Spurring their horses, Loving and Wilson rode hard for the Pecos and reached it ahead of their yelling pursuers. Their mounts were killed, but the two men scrambled down the steep cliffs of the canyon and hid in the high canebrakes. Shooting hard and fast they drove back the Indians. One, bolder than the rest, slipped close through a growth of polecat bush. As the Comanche rose to fire, Wilson killed him, but too late to save Loving. The savage's bullet broke the cowman's wrist and bored into his side. Then darkness descended to save the besieged white men.

They waited tensely until moonrise. Loving was bleeding badly and believed he was going to die. He pleaded with Wilson to save himself and at last Wilson agreed, in the hope that he could find help and return. Removing his clothing except for his underwear and shirt, the one-armed cowboy tried to swim the Pecos. His rifle, clutched in his single hand almost drowned him, and finally he left it on the bottom and floated down the current. Once he was almost discovered by a Comanche on the river bank, but the fugitive hid under some smartweeds and escaped. Eventually he was able to wade ashore and seek the trail herd.

Shoeless and nearly naked, Wilson walked almost three days across the cactus flats before he saw the dust of Goodnight and the cattle. Within a few minutes after he croaked out his story, a rescue party was riding for the place where Loving had been left, but the cattleman was gone. Two weeks later Goodnight learned from Jim Burleson, whose outfit had preceded him, that Loving dragged himself five miles after Wilson departed, and hid in a narrow gap. He was without food five days. Some Mexicans at last found him and for $150 took him to Fort Sumner.

When Goodnight reached the fort, Loving was up and around, but a relapse occurred owing to infection of the wound, and he died. On his death bed he asked his partner to continue the partnership until there was enough money to care for his family. This Goodnight faithfully did. By the next year he had cleared $80,000. Of this he turned $40,000 over to Mrs. Loving, bought out his partner, J. W. Sheek, for $20,000, and with the other $20,000 and the cattle of the C V brand, went into business for himself.

The next year he took up a new enterprise. Plenty of herds now were coming over the Horsehead route and Goodnight contracted with the famous John Chisum to bring over cattle to him on the Pecos, paying Chisum one dollar above the Texas price. In seven years Goodnight bought ten thousand cattle from Chisum. But the trail drivers saw another opportunity. Gathering a huge herd of ten thousand cattle Chisum moved over into the Pecos valley himself. They still talk of that immense cattle migration. The herd strung out for miles on the trail and at night when it was bedded down it required the night riders three hours to walk their horses around it. Chisum took his cattle to the Bosque Grande, where he settled and established the Long Rail and Jingle Bob brand. Also out of that settlement arose the Lincoln County war and Billy the Kid, of which more later.

Other cowmen followed the example of Chisum and Goodnight. The trend toward the Southwest was on, and as fast as the soldiers could corral the Apaches, New Mexico and Arizona were taken up for cattle range.

One more trail-driving exploit of Goodnight's deserves recording. The year after his partner's death he took a herd north and found tough old Uncle Dick Wootton still guarding the Raton Pass and demanding ten cents a head for cattle passing over it.

"Go to hell," Goodnight told him. He turned his cattle east and blazed a new trail of his own across Trincheras Pass which actually cut one hundred miles off the distance to Pueblo. Up this he took his cattle and sold them to John Iliff on the Platte. It was the trail which eventually pushed north to Cheyenne and the Chugwater valley of Wyoming, and it practically ruined Uncle Dick's easy toll-road sinecure over the Raton Mountains.

3. . . CAPITALS OF CATTLE LAND

XIV

"TO SEE SOMP'N HAPPEN"

The first days on the trail were filled with incident and excitement. Not only was it always a tense business to start a herd away from its home range, but the very adjustment to routine was full of interest and the cowboys were busy becoming acquainted with one another, growing accustomed to new names and faces, tentatively exploring, in their reserved, inarticulate manner, new personalities.

There was much singing and skylarking in the preliminary stages of the drive. The air was like wine, the long, horned procession with its medley of cattle sounds, was inspiriting to watch. Youth was in the saddle, action was in the atmosphere, and there was promise of high adventure, with far lands and strange sights to be seen.

All this was sufficient, for a time, to invest the camp and the daily drive with joyful lightheartedness. Around the chuckwagon fire at night cowboys exchanged experiences and told yarns. Usually an outfit contained one or two singers, more or less gifted according to the standards of their class, who possessed a repertoire of the songs of cattle land. "The Dying Cowboy," "Dinah Had a Wooden Leg," "Hell Among the Yearlin's," "Oh, Susannah," "When the Work's All Done Next Fall," and other favorites were done over and over, the cowboy chorus coming in to aid the soloist. Mouth harps were frequent adjuncts, and here and there a cowboy fiddler tucked his precious instrument among the dunnage in the chuckwagon. Good-humored pranks were played—always good-humored, for there was no telling how a really spiteful act might turn out, and politeness is a lesson well learned where youth goes always with a weapon strapped to its waist. A pack or two of cards appeared, and poker, monte and seven-up had their innings, although the stakes were microscopically small, since the cowboy who carried any appreciable sum of money on the trail was almost unknown. Even night herding was fun in those first days out.

But as time passed the drive became less amusing. One river crossing is very much like another river crossing, and one day's drive varies little from the next. Moreover, the discomfort of riding all day in rain, and trying at night to sleep with a river running down your spine, quickly palls. By imperceptible degrees the jubilant atmosphere departed. The stories soon were told until everyone was familiar with them; the very songs grew tiresome, since they must be listened to as rendered by the night herd each night and all night. Worn and greasy and torn grew the decks of cards and playing for a tobacco plug or a few cents as a stake lost its savor. Men became silent and withdrew into themselves. Always the same faces were about. From the first the food was monotonous. Each day was exactly like the last, until every man in the outfit, while dreading the crisis of a sudden night stampede, or the peril of swimming an angry river with the herd, almost welcomed them because they broke the dreary round.

A month passed; two. Sometimes three or four months, and still the cowboys were bound by an unbreakable and invisible chain to their herd of cattle, for there was one invincible code of the cow country—you remained faithful to the death to your outfit as long as you were on duty. A man's moral obligation to the herd was as great as that of a sailor who has signed articles for a voyage.

Then, one day, a distant sound came to the ears of a cowboy, perhaps riding point far up in the front. He raised his head and listened intently. It came again—perhaps the pin-point crow of a rooster far away, or even the barking of a dog, attenuated by its remoteness. The cowboy reined back his horse until he fell in with the rider next behind him. Possibly the two had been aloof for weeks, with scarcely a word between them. Now the first rider remarked with expressionless face: "Listen, Hank. I thought I heard a chicken crow jest back."

The second cowboy listened. Again came the far-away sound. There was instant brightening in the intent features.

Word was passed to others. A change in the atmosphere became immediately noticeable. New alertness imbued each rider. Faces relaxed into humorous lines. Quips flew back and forth. Even the horses pricked their ears, and the cattle, sensing a different spirit, stepped forward more briskly to their destiny. They were approaching civilization at last. The faint sound was a message from some nester's home, down in a valley somewhere, and it meant the end of the trail was at hand—the cattle market was almost reached.

A day or so later, if the herd was sold and delivered promptly, the cowboys would be paid off. With bridle reins loose and spurs plying they would race, yipping, for "town," their hot young blood starved for the only break they knew in the unbearable monotony of their lives— liquor, gambling, women, strong excitement of any kind.

And "town" would be ready to furnish them—at a price—
that excitement.

Fermenting, festering, boiling over with energy and evil,
the trail towns sprawled across the West. Baxter Springs,
Abilene, Ellsworth, Newton, Wichita, Caldwell, Hays
City, Dodge City, Ogallala, Cheyenne, Tascosa, Roswell,
Lincoln, Miles City, Denver—every one of those names
and many others rang through the saga of their era. One
after another, for greater or lesser periods, they became the
foci of the cattle trade. And during its own hectic years
each saw its share of excitement and tragedy, of sordid
crime and immorality, of gambling for great stakes in
money or lives, of heroism and cowardice and of a keyed-
up, dangerous existence.

Baxter Springs was first. Circumstances and conditions
made it a cowboy center for one summer—1866—when a
quarter of a million cattle pushed up from Texas to the
Missouri-Kansas border and there found a barrier of resolute armed farmers, abetted by the bird-of-prey Jayhawkers. For months the herds jammed the bedding grounds
south of Baxter Springs while their owners tried vainly to
solve the impasse and cowboys and *vaqueros* idled. Baxter
Springs opened itself to the idle population and made shift
to capitalize on the desire of Texans "to see somp'n happen." Down from the Missouri River towns came a scab-
rous horde of exploiters. Saloons opened to augment those
already present. Gambling halls sprang into being. The
first of a long line of red light districts, the inevitable con-
comitant of every cattle town, developed. For that summer's brief months Baxter Springs was the liveliest place in
the West, but its career ended then, too short to earn it a
reputation to compare with its successors. In the fall the
cowboys departed and next year the combination of grangers and the opening of a market at Abilene put Baxter
Springs forever in eclipse.

Of the doings at Abilene there is ample record and ample

controversy. There, as before, the cattleman came squarely into collision with the farmer. Here was a contest as old as civilization and as bitter as the inborn animosities of two clashing ways of life. Two races, in effect, had their meeting. The settlers around Abilene were from the North— farmers, land speculators, money lenders, merchants. They were strong in respectability, a virtue ingrained with the New England tradition of piety combined with sharp dealing, which did not impeach a man's high standing as a pillar of the church merely because he might defraud his neighbor. They were dogged, persistent folk, rooted strongly to the ground, people of the plow and the counting house, who measured their worth in the dollars they possessed and the acres they owned, great upholders of the letter of the law.

Across their path now rode the cowboys, a unique cavalry, weaponed, humorous, unstable, sometimes murderous, the horsemen of the South. Theirs was the tradition of the cavaliers and the background of a free and easy mode of living where one man was as good as another regardless of the number of cattle under his brand. As for the desire to own the soil, that was beyond the comprehension of many of them. Land in Texas spread for uncounted leagues—so much of it that it had no value. It was free to any who wished to use it, and that anyone should build a fence around a little corner of it, and exclude the world from it, and grub in it for a living, was repugnant. There was, moreover, a deep suspicion, among these riders, of written law and lawyers. Books they knew little, but they had a code, and a man made good his slightest word or he was no man at all among them.

"I've seen many a transaction in steers running as high as five thousand head and involving more than $100,000, closed and carried out to the letter, with no semblance of a written contract," G. W. Rourke, who was a railroad agent at Dodge City, Kiowa, and other cow towns during the great period of the open range, once told me. "Those

Texas cowmen had a regard for their word so meticulous that it was rare to see a paper exchanged between them in business. A nod of the head was enough. They never questioned the integrity of one another."

Even this simple, almost childlike code of honor was to disappear in the succeeding years as speculation and big money from the East and foreign countries entered the range, but in the meantime it was symptomatic of ideas which were divergent from the careful notions of the settlers. The clash between these conflicting forces was to be eventful, but it was to end inevitably as have so many clashes between swirling clouds of cavalry and solid phalanxes of footmen—in retreat for the cavalry. Abilene was the first great battle ground.

The circumstances under which Abilene sprang into prominence have been recited. It remains to be told what happened to Abilene when Joseph McCoy concentrated in it the Texas cattle trade and the cowboys tramped through the streets in their narrow, high-heeled boots.

Startlingly apparent was the boom in the town, as soon as the first trail herds arrived. Where a handful of log huts had stood, frame buildings began to rise. By 1870 there were no less than four hotels, ten boarding houses, five general stores, and ten saloons, besides other establishments, some of which were of a type distinctly unconventional. With few exceptions these structures were of one story and they affected that peculiar style of architecture which was to place its imprint on the West—the false front. A high board façade, seeking to give the illusion that the building was of two stories—and fooling nobody—is a transparent and shoddy device observable still today in some Western towns. In the early days it was as typical of the cow towns as were the great heaps of tin cans—heaps which, rusting away, are today sometimes the sole markers of once-vivid communities, long since gone.

South of Abilene's railroad track was "Texas Town," a

zigzag thoroughfare, flanked by saloons, gambling houses and bagnios. Famous saloons stood there, the Bull's Head, run by Phil Coe and Ben Thompson, the Applejack, the Old Fruit, and the Alamo, the latter notable for its three sets of double-glass doors—unimaginable magnificence—and the fact that it was the headquarters of Wild Bill Hickok when he was city marshal. The Drover's Cottage, built by McCoy and run by J. W. Gore and his wife—known to everyone as Lou—was the swank hotel where the wealthy ranchers lived while in Abilene. It was three stories high, contained one hundred rooms, and had a barn for one hundred horses and fifty carriages. Most of the ordinary cowboys camped outside of town or lived in one of the lesser hotels or boarding houses.

At first no attempt was made to control the wild Texans. They stalked down the streets, the rowels of their spurs whizzing, and the fringes of their chaps swinging with the swing of their holstered revolvers—hunting for amusement. Stuart Henry, who was a boy in Abilene when his brother, T. C. Henry, was its first mayor, tells in his book *Conquering Our Great American Plains* how the citizens viewed the cattlemen with a mixture of scorn and alarm, and yet were not above profiting from them. The Texans shambled in their walk from much riding, wore spectacular costumes, drank their liquor straight, and were prone to settle any dispute with a flurry of revolver shots. Such visitors made the good citizens of Abilene extremely nervous. No telling when one of them might go on the rampage, shoot up the town, and perhaps pot a couple of deacons. Yet the apprehension seems ill-founded. Nobody was killed in 1869 or in 1870. In fact, nobody was killed until the advent of officers of the law, employed to prevent killings.

A step toward "law and order" was taken in September, 1869, when the town was incorporated as a third-class city, with a board of trustees made up of Henry, McCoy, Hersey and Thomas Sheran. Henry was chosen chairman, which made him the first mayor of Abilene. Straightway action was taken to avert the perils which everybody feared for the town. An ordinance was solemnly promulgated making it illegal to carry a deadly weapon inside the city limits. Notice to this effect was posted in conspicuous places, but it had the defect of quickly becoming illegible. The cowboys persisted in punctuating the signs with their revolvers and the bullets obliterated the legend. As for abolishing the weapons, it had no effect whatever.

As if this flouting of authority were not enough, a worse thing happened. The city fathers created the office of marshal and began looking around for a man to fill the post. Meantime the erection of a small stone jail was begun in the middle of Texas Street. On this construction project, feeling perhaps that personal implications were involved, the Texans looked with animosity. One night they demolished it.

This time Abilene's righteous citizens were thoroughly affronted. A guard was placed and, with armed men patrolling day and night, the walls of the jail were rebuilt and the roof securely bolted on. Almost immediately it had an occupant—the Negro cook of one of the trail outfits, who had imbibed too freely and was amiably but unsteadily testing his marksmanship on the street lamps. Hardly was he incarcerated, however, when Texas Town erupted. Throw their nigger in jail? The time for patience was at an end. A band of cowboys rode up to the little prison early in the morning, blew off the lock with a six-shooter, and the Negro was free. Then the Texans, giving vent to what some Abilene chroniclers have described as "the Rebel yell," but which in reality was probably nothing more than the cowboy whoop of exuberance and defiance, closed up all business houses in town, rode into the saloons on horseback, and finally galloped gleefully out into the prairie.

Excited citizens gathered, formed a posse, and went dashing forth in buggies to recapture the colored prisoner. At home their women apprehensively barred their doors, expecting nothing less than outrage from the cowboys while the men were charging off to duty. In this, Abilene's prevailing notion of the Texas horsemen reached the limit of absurdity. Anyone who knows the exaggerated respect the cattle country has always paid womanhood knows the women of Abilene were in no danger of rape. It was a part of the hysteria of the day.

Weary and dust-covered, the posse struggled back late that evening in its buggies. The Negro was not brought back, although some cowboys had been encountered on the trail and returned to town under heavy guard. It was assumed that they were "accomplices" of the Negro, although this batch of Texans, plainly astonished by their treatment, seemed to know nothing of the day's exciting events. In the end the trustees voted down an hysterical and bloodthirsty motion to hang the prisoners. They were released to return to their camps.

The jail delivery was the final, ultimate straw, and the town trustees with earnestness set out to find a marshal. A man named Smith applied for the job, but he bore the name of "Bear River Tom" Smith, and the reputation of having been a leader in a famous mining country battle. The worthy citizens of Abilene thought him a little too ferocious. Two soldiers with high recommendations came from St. Louis to look over the situation. The cowboys, learning of the visit, put on for them a special exhibition of choice and selected deviltries. Without stopping for a formal farewell, the soldiers departed on the next train.

While the citizens, with the watchword "Purity and Peace," trembled in their homes and stores in fear of a Texas attack, Mayor Henry in despair sent telegrams westward to trace Smith. "Bear River" or no, Abilene had to have someone to quell its disturbances. The righteous townspeople, as is always the case with the timorous, anticipated consequences far worse than the reality.

In actuality the Texans, encamped out on the prairie, had no mass resentment against Abilene and its puritans. To them the town was a place merely for a few days' spree. They were constantly going and coming, new groups bringing their herds; and after a day or two of carousing, departing quickly southward over the Chisholm trail for Texas. There could be no continuous, motivated hostility in such a changing population. A fundamental difference in viewpoints did exist which made the Texans contemptuous and even hostile toward the money-grubbing citizenry. Always men who hold life lightly and risk it often, and who care little for pampering their bodies, have such contempt for other men who are preoccupied with safety, are easily frightened, and who fear hardship. Outside of perhaps some insolence, however, most of the resentment of the Texans displayed itself in a sort of humorous perversity—and the fact remains that no serious harm was done in Abilene.

XV

MARSHALS OF THE PEACE

Thomas J. Smith, known as "Bear River," broad-shoul-dered, erect, trim-hipped, one hundred and seventy pounds of muscular effectiveness, with wide Irish gray-blue eyes, a well-kept mustache and thick auburn hair, arrived in Abilene on July 4, 1870, in response to Mayor Henry's telegrams. At his handsome and athletic appearance the timid hearts of Abilene acquired confidence. Much later it was learned that he was a trained police officer, having re-ceived his schooling in the New York department.

No sooner was his commission signed than the new mar-shal set about the enforcement of the abortive firearms or-dinance. Store keepers, saloon men and hotel proprietors were persuaded to post signs reading: ALL FIREARMS ARE EXPECTED TO BE DEPOSITED WITH THE PROPRIETOR.

Then, on his gray horse Silverheels, the marshal began patrolling the streets. His first showdown came the very Saturday night after he took office. "Big Hank," a cowboy with apparently no surname, who had previously defied au-thority, met the officer and abused him publicly, at the same time daring him to try to take the Texan's weapons. Big Hank was a gun-fighter, brave enough, and waiting for the marshal to draw, but he was unprepared for Smith's tactics. Without touching his own weapons, the officer suddenly stepped forward and swung his fist against the point of the cowboy's chin. It must have been a magnificent wallop. Down went the Texan, knocked out. When he recovered his senses, Smith stood over him with the cowboy's pistol, ordering him to leave town. The crestfallen cowpuncher obeyed.

That short, decisive clash became a sensational topic of

conversation in the cattle camps on the prairie about the city. Fist-fighting was a new, unorthodox thing to the Southerners. In Dixie the idea of pugilism was contrary to a gentleman's code. Sometimes two Negroes were pitted against one another, much as might be two cocks or two bulls, but white men disdained to come to blows. If the argument was important enough to warrant a personal encounter, it was important enough to invoke the *code duello*, either formally or informally. A gentleman guarded his honor with his life. Death in such a cause was noble and dignified—but there was nothing noble or dignified about a bloody nose or a black eye. Of course, the Texas cattle frontier was uncouth and rough, and all this was not articulated, perhaps not even realized. But it was the tradition back of the Texan's mode of conduct.

Yet there was ungrudging admiration for Marshal Smith, too. It took guts, the Texans agreed, to go up against a man all ready to pull a hog-leg, and knock him down with a fist. To a race unacquainted with boxing, moreover, there was something uncanny about the mere feat of felling and stunning a man with a blow. It appeared superhuman and awesome. To Smith, the son of Irish immigrants, brought up in the streets of New York, boxing was a familiar technique and knocking out a man a matter of calculated science.

"Bear River Tom" increased immensely in public stature through his exploit, but once more he was challenged before he obtained full sway. This time it was one Wyoming Frank (even in that day the West was chary of furnishing last names and complete data on careers—for excellent reasons in some cases). The cowboy, camped on Chapman Creek northeast of Abilene, made a bet that the marshal could not "bulldoze" him. It was a bad bet. Smith backed Wyoming Frank into a saloon and stepped in with the deft movements of a trained boxer. Awed witnesses later described it as a "double blow." Probably the marshal used what Jack Dempsey has called the "one-two"—a jolt to

the midriff to bring the guard down and a terrific uppercut to the unprotected chin. Wyoming Frank recovered
consciousness sometime later, greatly subdued.

Abilene was thoroughly tamed. No Texan wished to
risk meeting those pile-driving fists. There was no defense
against them unless a man deliberately shot down the marshal without waiting for him to draw. And that would
have been murder and the guilty person liable to a "neck-
tie party," perhaps at the hands of his own associates, for
the cattle country had strong ideas on the matter. Thereafter the cowboys turned in their weapons when they
came to town. Abilene citizens breathed a sigh of relief.

Then, unexpectedly, Marshal Smith was killed—not by
Texas desperadoes, but by Abilene settlers. After the cowboys went south that fall, the marshal, on November 2,
went to a dugout a few miles northeast of town to arrest
Andrew McConnell for the killing of John Shea in a
boundary dispute. McConnell and his friend, Moses Miles,
a Massachusetts Yankee, refused to surrender. They drove
off James H. McDonald, the marshal's companion, and
killed Smith, first shooting him, then nearly severing his
head with an ax. A posse caught the culprits, but neglected
to lynch them. Whereat they sought trial in another county
and received only penitentiary sentences for their crime.

So the Abilene sheep once more were without a watch
dog to guard them. The sheep bleated loudly and the city
fathers looked about for a successor to Bear River Tom.
Abilene sank daily deeper into a state of sin. In 1871, the
height of the cattle trade, and following the death of
Smith, John Wesley Hardin, a Texan who is credited with
killing forty men and was therefore no soft-boiled egg
himself, had this to say of the place:

"I have seen many fast towns, but I think Abilene beats
them all. The town was filled with sporting men and
women, gamblers, cowboys, desperadoes and the like. It

was well supplied with bar rooms, hotels, barber shops and gambling houses, and everything was open."

This is high tribute from so eminent an authority on sinfulness as John Wesley Hardin. Observe that this un-curried son of Texas listed barber shops in the category of presumably forbidden luxuries along with bar rooms and gambling houses. Perhaps he considered them a symptom of degenerate effeteness.

Other men of evil eminence almost equal to Hardin's came to Abilene that year of 1871. Ben Thompson, a man with a sinister succession of killings, and his younger brother Billy, also an aspiring gunman, were two. Ben went into partnership with Phil Coe. Both were Confed-erate veterans and killers. They bought the Bull's Head saloon and gambling hall.

Just before the cattle season opened, still another illustri-ous virtuoso of the six-shooter arrived. But this one came on the side of the law. The Abilene trustees decided to do the big thing for the town and imported the most famous gunman of his day, Wild Bill Hickok, to fill the vacant post of marshal. Christened James Butler Hickok, Wild Bill is said to have compiled, before his death, the longest authentic string of killings in the history of the West— some seventy of them. He was a tall, graceful, pantherish man, almost as vain about his looks as a woman, a great gambler, and possessed of tremendous courage. His most notable quality, however, was the almost miraculous speed with which he could draw his revolvers and the unbeliev-able accuracy with which he fired them. He was a two-gun man and, for all the literature, ambidexterity with six-shooters was rare and always will be.

Unlike his predecessor, Wild Bill depended upon his weapons and his reputation. Although he was probably several degrees more deadly than Tom Smith, the event was to prove that the unusual tactics of bare knuckles were superior, in Abilene's case at least, to the hot lead of the

most respected killer of the West. The cowboys understood gun-play. Although it meant death to the loser, they
did not dread it as they dreaded the ridicule which might
follow an uppercut from the New York Irishman.

Wild Bill served conscientiously, according to his standards. He made his headquarters in the Alamo saloon instead
of the city hall, because there he could augment his salary
with gambling winnings. But he carried out some difficult
and dangerous duties. Once he was required to bring a
refractory trustee to a trustee meeting to form a quorum.
Hickok went to the man's office, threw him over a shoulder, and dumped him in a chair at the council table. Another time he was ordered to close all dance houses and
stop the vending of intoxicants on Texas Street. Once he
was asked to clean up the red light district and run all the
prostitutes and gamblers out of town. Again, to institute
a general cleanup of undesirable elements in the city. Most
of these were perilous assignments, but the marshal's reputation was so ferocious that he carried them out without
opposition.

Yet no longer did the Texas cowboys turn in their
weapons when they rode into town. Perhaps this was due
to a tolerant attitude on the part of Hickok, who no doubt
felt that a brace of six-shooters was a legitimate feature
of any man's costume. There was some shooting. Finally
enmity developed between the owners of the Bull's Head
saloon and the marshal, ending in blood.

Ben Thompson was in Kansas City when the feud started
between his partner Phil Coe and Wild Bill. Cards seem
to have been the genesis of the trouble. Hickok is said to
have caught Coe cheating and "called" him sharply. Coe
did not dare to fight, but he did not forget. Later he and
Thompson charged that Wild Bill was permitting crooks
to fleece the cowboys, for a share of the spoils, and the
Bull's Head proprietors adopted the very moral attitude of
opposition to such practices. A second story, probably

attributable also to the partners, was that Coe had bested Hickok in rivalry for the favors of one Jessie Hazel, a fair but frail dance hall habitué, and Hickok vowed he would finish his rival.

Whatever the cause of the quarrel, it came to a climax the night of October 5. Preparing to go south for the winter with some of his Texas friends, Coe was holding a final celebration. The revelers came upon Wild Bill and demanded that he join them. He refused, but invited them all to take a drink with him. Later the roisterers left the saloon and trooped noisily and drunkenly down the street. A dog came running down an alley and Coe exuberantly took a shot at it, killing it. The shot brought Wild Bill running.

In the street stood Coe, revolver in hand.

"Who fired that shot?" snapped the marshal.

"I did!" Coe answered belligerently.

No question of explanations here; both men knew the climactic moment had come. Like the flicker of light, Hickok's hands shot to his holsters and both guns flashed out. Simultaneously Coe began firing. His shots went wide. The marshal's first bullet smashed through Coe's abdomen, and Hickok holstered his pistols remarking: "I've shot too low."

At that moment another man came running around a corner. In the darkness Wild Bill could not tell who the newcomer was, but he was alone in a crowd of enemies and he supposed it was another foe. He drew and shot twice. The man fell. He proved to be Mike Williams, Hickok's own deputy and friend.

Stricken with grief at the tragedy and berserk with anger, Wild Bill put an instant end to the festivities and everybody was ordered from the street. The following Sunday Coe died. It was the last shooting of the kind in Abilene. Additional deputies were sworn in by Hickok who began carrying a sawed-off shotgun, loaded with buckshot. Now, when it was too late, he enforced the deadly weapon ordinance and did so until, a week later, he was relieved of his commission.

There was no further need for him anyway. The cattle trade at Abilene was gone. Next year it shifted to other towns and the sheep were safe without a watch dog.

XVI

THE WILDNESS OF ELLSWORTH

After the citizens of Abilene addressed their top-lofty proclamation to the Texas trail drivers inviting them to stay away, the cattle business which had built the town was cut off as if by a bowie knife. This was not entirely because of the proclamation, although that helped. Already the railroads were spreading further toward the cow country and Abilene would in any case have lost the trade eventually.

Among the inheritors was Ellsworth, Kansas. It started as a rival to Abilene, became a successor, and finally lapsed into obscurity even greater than that of McCoy's first adopted city. In 1871, 35,000 cattle were shipped from Ellsworth's stockyards over the Kansas Pacific. The following year—Abilene's last—the total was 40,000. In 1873, when Abilene was eclipsed completely, twenty-eight herds, ranging from two thousand to ten thousand head each, were pointed for Ellsworth as early as April. Colonel J. J. Meyers alone trailed 27,000 cattle to the town late in the season, the number being broken up, of course, into several herds. In June the Ellsworth *Reporter* estimated more than 100,000 longhorns were being held about Ellsworth and that 250,000 would be received and shipped before the end of the season. The town boasted that its stockyards —built of unpainted lumber—were the biggest in the state. They covered several acres and possessed no less than seven chutes whence two hundred carloads of cattle a day could be loaded. Colonel Robert D. Hunter, who was later to win some reputation for his "legal rustle" of John Chisum's cows, was the manager in behalf of the Kansas Pacific.

From the first Ellsworth had a history more lurid than

the more publicized Abilene. The town was laid out in
the spring of 1867, a typical townsite boom project; but
a flood by the Smoky Hill River turned the site into a
lake and washed some of the flimsy houses from their
weak foundations that very June. There had been about
two thousand people in the town; this population now
shrank to about fifty, and the townsite was moved to
higher and safer ground. By 1871 Ellsworth had been incorporated as a third-class city, had elected a mayor and
city council—and appointed a marshal. It was time; the
cattle trade was beginning to arrive.

At the height of the brief livestock boom a thousand
people constituted the town's permanent population, but
this was probably doubled by an additional, impermanent
population—a typical cow town collection of saloon keepers, gamblers, harlots, cattle traders, sharpers, gunmen,
cowboys, and thieves. For a time the legitimate business
men were in the minority.

In addition to its stockyards Ellsworth had one civic
boast of which it was immensely proud—it had a sidewalk.
This was no wooden board walk such as other Western
towns may have had, but a solid slab of magnesia limestone.
To be sure it was not extensive—only twelve feet wide,
the length of the Grand Central Hotel—but it was the only
sidewalk, legitimate and genuine, west of Kansas City in
the plains country. Arthur Larkin, proprietor of the Grand
Central, profited by that enterprising sidewalk. His house
became the chief headquarters for Texas cattlemen.

Through the center of Ellsworth ran the Kansas Pacific
tracks; the business section stretched for three or four
blocks along both sides of the right-of-way, giving the
appearance of an exceptionally wide street. Ordinarily this
highway was referred to as two streets, North Main and
South Main, according to which side of the railroad the
speaker was discussing. Besides the hotel with the notable
sidewalk, the Drover's Cottage stood on South Main, operated by J. W. and Lou Gore, the former managers of the

hotel of the same name in Abilene. It is said, in fact, that part of the Abilene building was hauled over to Ellsworth and incorporated into the structure there. There were one or two other hotels of lesser fame.

Somewhat separated from Ellsworth proper and standing half a mile to the east, was an unclean, unwholesome community of shacks and jerry-built houses known as Nauchville. It contained most of the brothels, together with the inevitable accompanying saloons and gambling joints. On the flats near at hand was a horse-racing track. Most of the evil which sprouted like blotched fungus in Ellsworth had its roots in Nauchville.

As early as July, 1872, a couple of gamblers, Kennedy and Olive, shot it out in an argument over cards. Both were wounded—Olive dangerously—but both recovered, and Kennedy, arrested as the aggressor, escaped from jail and departed for unknown lands. But the really sinister element did not begin to arrive until the spring of 1873. Then Ben and Billy Thompson, the old enemies of Wild Bill Hickok, came down from Abilene. Shortly after that Cad Pierce and Neil Cain, a pair of cattlemen, wild and fierce, friends of the Thompsons, came up the trail from Austin, Texas, with a herd of cattle. Another addition was John Sterling, a gambler with an uncertain disposition. There were many more of the same class.

As the influx of Texans began the city council passed an ordinance prohibiting the carrying of firearms, similar to Abilene's, and appointed a police force. This, in 1873, was composed of Marshal John W. Norton, and Deputies John DeLong, John Morco, John S. Brautham and Ed Hogue. The array of Johns was enough to confuse anybody, and for the sake of differentiation the four Johns on the force were known, in the above order, as Brocky Jack, Long Jack, Happy Jack, and High-Low Jack. Happy Jack Morco owned a bad reputation, claiming to have killed twelve men on the Pacific coast. An unfortunate plan had been adopted by the city, whereby deputies received as fees a portion of the fines collected for violations of laws, and Happy Jack, trying to increase his fees, became so insufferably officious that his activities possibly had some bearing on the subsequent troubles.

First of the Texans to have contact with the court presided over by Judge Vincent B. Osborne was Ben Thompson himself. One night, according to an old Ellsworth

story, he became drunk and created a disturbance. Nobody attempted to arrest him because of his dangerous reputation, but next day he appeared before Judge Osborne of his own volition, solemnly swore out a complaint against himself, entered a plea of guilty to the charges, listened to the judge assess a fine against him, and paid the fine.

In spite of this exemplary conduct on one occasion, the Thompsons were turbulent and feared. One of the few men in Ellsworth who could handle them was Sheriff Chauncey B. Whitney, a brave and honest officer, who had been one of the fifty-two men with Major G. W. Forsyth in the celebrated Battle of Beecher's Island against the Cheyennes in 1868. Both Thompsons liked and admired Whitney, and because of their friendship for him respected his wishes.

A big game was going on in Joe Brennan's saloon on South Main on the morning of August 15, 1873. Neil Cain was dealing monte, and Cad Pierce, flush after having sold his cattle, was betting heavily, insisting on higher and higher stakes, to which Cain objected. Near the table, watching, stood Ben and Billy Thompson, both drinking heavily and in the condition when they became dangerous. Failing to induce Cain to increase his limit, Pierce called to Ben Thompson:

"Hey, Ben. Can't you find somebody to take my over-bets? This here puny limit is making me some weary."

Thompson summoned John Sterling, a case-hardened gambler, who would take a chance on anything, and who agreed to take all bets above the sums wanted by Cain. According to Thompson he said, as he entered the game, "If I win, consider yourself one-half in." If Sterling did make that statement the situation was unusual, because there is no evidence that Thompson staked him or did anything more than invite him over.

Sterling won heavily. Eventually he had more than a thousand dollars of Pierce's money. Pocketing this he started away. Thompson's eyes narrowed as he watched the gambler go, and the irritability of his conduct toward the other men about him became noticeable. He was brooding over the failure of Sterling to meet the obligation, real or fancied, which Thompson felt was owed him. Both Thompsons began drinking and their faces grew dark and dangerous.

Those who knew the Thompsons saw the storm signals flying and when word of the situation came to Sheriff Whitney he canceled a picnic excursion to which he was invited and elected to remain in town, hoping to avert trouble. The explosion was inevitable, however, and it came that same afternoon. In Nick Lentz' saloon where

Ben Thompson was drinking he suddenly found himself confronting Sterling. Both men by this time were drunk and quarrelsome.

"Where's the half of them winnings you promised me?" demanded Thompson. Some sharp words were exchanged. Then Sterling did something few would have dared to do. He walked forward and slapped Ben Thompson squarely across the face. Happy Jack Morco was in the saloon, and he hastily pulled his revolver and covered Thompson. But Ben simply blinked his eyes and made no move to resent the blow. It developed that he was unarmed, but Sterling could not have known that when he struck him. Later a good many men acknowledged that the gambler was possessed of uncommon courage, or foolhardiness, to have slapped a man like Ben Thompson. Ordinarily a man would not have been expected to survive more than a fraction of a second after the blow was struck.

Yet Sterling played in luck, or perhaps he was confident of his ability to match anything Thompson attempted— and Ben walked out of Lentz' place and back to Brennan's saloon where Cad Pierce, Neil Cain, Billy Thompson and other Texans were loafing. Scarcely had he entered the bar room when Sterling and Happy Jack appeared suddenly in front of the saloon, armed and yelling: "Get your guns, you damned Texans, and fight!"

In the saloon, men were diving for weapons or seeking cover behind the bar and any other obstructions, no matter how slight. Morco and Sterling did not enter, but retreated up the street.

"Give me a gun. Give me a gun," Ben Thompson was begging. Nobody obliged him and he ran over to Jake New's saloon where his weapons were. There he snatched up his revolver and Winchester rifle, and his brother Billy, following him, picked up a double-barreled shotgun loaded with buckshot. Drunk and careless, Billy cocked both barrels of his gun. As the brothers stepped out in front of the saloon, one of the barrels was accidentally discharged, the heavy load splattering with lead the sidewalk in front of the Grand Central Hotel, close to the feet of Major Seth Mabry and Captain Eugene Millett. Those gentlemen, two of the most important and wealthy of the Texas ranchers, had won their commissions fairly in the Confederate army, but that did not prevent their retreat from being speedy and undignified.

Across the street to the railroad track, where there was a clean sweep for the guns, walked the Thompson brothers, and shouted a challenge to Sterling and Morco to come out and fight. In the meantime someone had run for Sheriff Whitney, who came hurrying down the street at this minute. Relying on his friendship with the two men, he walked toward them to pacify them. Nobody else in Ellsworth could have done that, but Whitney talked to them and succeeded in inducing them to return to Brennan's.

They had almost reached the saloon when a new alarm came. "Look out, Ben!" yelled W. A. Langford, a Texan. "Here they come with guns!"

Into Beebe's general store leaped the Thompsons, followed by Sheriff Whitney. Happy Jack Morco appeared, drunk, armed, and aggressive. Ben Thompson's six-shooter barked but the quick shot missed. At the same instant a deafening roar came from the rear of the store. Billy Thompson's shotgun had gone off again.

The Thompsons always swore it was an accident: that Billy, drunk, had never thought to lower the hammer on the unfired barrel after almost killing Major Mabry and Captain Millett. If it was an accident it was a fatal one. The full charge struck Sheriff Whitney in the side, tearing a great hole there.

"My God, Billy, you've shot your best friend!" cried Ben.

Whitney screamed, "I'm shot!" Then he collapsed, gasping, "Send for my wife, I have received a bad shot."

The sheriff was carried to his home a couple of blocks away, and a surgeon was sent for at Fort Riley. But the buckshot had riddled Whitney's lungs and he died three days later.

Meantime Billy Thompson left Brennan's saloon through the rear door and mounted a horse. The Texans were true to their friendships to the last. Neil Cain brought the pony from a livery stable and held it while Billy mounted. Cad Pierce handed up a roll containing a hundred dollars in bills. "Here, Billy," he said. "You'll need this." Then Billy, according to the account by the local newspaper, "rode slowly out of town cursing and inviting a fight."

The failure of his police force to function made Mayor Miller of Ellsworth furious. He called in the entire group and discharged the men immediately. A half-hearted posse pursued Billy, hoping fervently it would not catch up with the straight-shooting and daredevil youngster. He escaped easily and was not captured until three years later, when he was arrested in Texas by the rangers who returned him to Ellsworth, where he was tried on the murder charge and acquitted by a jury which held that the killing of Sheriff Whitney was accidental.

Ben Thompson was arraigned the next day after the shooting on a charge of assault against Happy Jack Morco, but the case was dropped when the complainant refused to appear.

Ellsworth reorganized its law-enforcement force. Ed Hogue was made marshal and the citizens of the town began the secret organization of a vigilance committee. News of this soon reached the Texans. Five days after the Whitney shooting, Ben Thompson warned Cad Pierce that the vigilantes were after him. Immediately the Texan rode to where Hogue stood by Beebe's store and, dismounting from his horse, confronted the marshal with the demand to know if the story was true. Hogue denied it, but Ed Crawford, one of the new policemen, stepped forward and challenged Pierce. He said later that the Texan "put his hand behind his back." Crawford's shot passed through Pierce's side. Stumbling and bleeding, the wounded man ran into Beebe's, the policeman in pursuit, gun blazing. A second bullet shattered Pierce's arm and then the maddened Crawford caught up with him and spattered his brains on the floor with repeated blows of his six-shooter.

The crowd in the store turned Pierce over and then it was discovered that the dead man was unarmed. The incident began to take on an ugly look—the aspect of cold-blooded murder. Pierce was a gambler and a Texan, but a square man and well liked. Feeling against Crawford grew.

Meantime the cleanup was launched in earnest in Ellsworth. Ben Thompson took a train for Kansas City. Its services were openly offered to the mayor by the vigilance committee, and a squad of twenty armed citizens patrolled the streets nightly. Some unfragrant individuals were "run out of town," and the "regulators" invaded the Grand Central Hotel, Texas headquarters, and seized the weapons there. A telegram went from the Texans to Ben Thompson at Kansas City, telling him to bring on a new arsenal, but just when matters grew most tense, the governor of Kansas took cognizance of the situation and offered Mayor Miller the assistance of the state armed forces to maintain order. Ben Thompson took another route back to Texas.

Happy Jack Morco, who left town after being discharged from the police department, returned to Ellsworth early in September. He was armed and looking for trouble when, on September 4, he met J. C. Brown, one of the new policemen. Brown sought to pacify Morco.

"You know it's against the law to carry weapons," he said. "Give me your six-shooter, Happy Jack."

Instead of complying with the friendly request, Morco drew his gun. Brown's eyes narrowed and his pistol roared twice. The first bullet passed through Happy Jack's heart; the second through his head as he fell.

It was November when the final act of Ellsworth's

bloody melodrama was played out. Because of the feeling against him, Ed Crawford, slayer of Cad Pierce, had departed from town. He was warned against returning, but early in November he was back again in Ellsworth, spending several days at his usual loafing places. A Friday evening came, and Crawford, with two friends, Charlie Seward and J. W. Noel, embarked on a tour of Nauchville, going from saloon to resort house, to dance hall, in a drunken round.

At last the trio of inebriates arrived at Lizzy Palmer's place, one of the best-known sporting houses. As the men entered they recognized some Texas cowboys who already were present. Everyone was drinking freely and presently voices were raised. One of them was Crawford's. He was in a fierce altercation with one of the Texans. Still shrill and loud-voiced with wrath, the two men stepped into the hall outside the room where the "guests" and their girls were drinking. White-hot flashes lit up the dark passage and the narrow space was choked with powder smoke as six-shooters were emptied. Screams from the girls and shouts from the men added to the bedlam.

Somebody carried a light to the hall. Slumped forward on his face on the floor lay Crawford. There was a bullet hole through his body and another through his head.

To this day it is not definitely known who killed Crawford, but most old-timers believe the slayer was a cowboy named Putnam, a Texan and a brother-in-law of Cad Pierce, who thus closed the vendetta for the family.

That ended the wildness of Ellsworth. With 1873 the cattle trade was gone.

XVII

NEWTON'S GENERAL MASSACRE

For one single, hectic season—from the time the Santa Fé railroad completed its line to the townsite in July, 1871, until the tracks pushed on to Wichita—Newton, Kansas, was the toughest, loudest, most dangerous spot in the West. The late Cal Johnston, who lived there from the day the town opened until his death recently, once told me, "The firing of guns in and around town was so continuous that it reminded me of a Fourth of July celebration from daylight to midnight. There was shooting when I got up and when I went to bed." How many men were killed in that brief reign of terror nobody will ever exactly know.

In less than three months Newton sprang into mushroom being. A line of flimsy buildings, mostly of the familiar "false front" architecture of the West, confronted each side of Main Street. About one in four of these was a saloon or dance hall. "The Mint," "The Side-Track," "Alamo," "Bull's Head," "Legal Tender," and "Gold Room," were a few of the euphonious titles of these establishments where gamblers, liquor dealers and prostitutes collaborated in abstracting cash from the wild youths of the cattle trail. Judge R. W. P. Rust, who visited Newton in its first days, gave the following description of one of the frontier dance halls:

"All night the hall was filled with cowboys, gamblers and roughs, who, in company with the 'soiled doves,' treaded the mazes of the giddy waltz until daylight came, or weary Nature failed to furnish strength. The building was crowded and drinking whiskey and gambling were going on at the bar or around the small tables in front; singing and preaching from a platform in the rear of the building; while in another room were scarlet women peering through the curtains at the scenes in front. After the religious services were over, the reverend gentleman who had conducted them was given a five-dollar bill, and induced to take a social glass, and promised to call again. Mr. Peter (one of the spectators) afterward told me that it beat anything he ever saw."

That was Newton in 1871. The railroad had moved down to it, and Joseph P. McCoy, having had a falling out with the Kansas Pacific and Abilene, came down and opened his second stockyards there. Although its connection with the Texas cattle trade was shorter than Abilene's, or even Ellsworth's, it eclipsed those towns in its record of blood.

Two Texas cowboys, Snyder and Welsh, inaugurated the summer's killings. They came up the Chisholm trail together with a herd of longhorns. On June 16 they were quenching a three months' thirst which they had built up on the long dusty trek, when a quarrel flashed out. Nobody knows the cause of the dispute; perhaps it was animosity built up during the long monotony of the trail. But in front of Gregory's saloon Welsh crippled Snyder and Snyder killed Welsh, in a revolver duel which gave to Newton's "Boot Foot Hill," soon to be plentifully populated, its first occupant.

A few days later a mysterious individual named Irwin was shot in the Parlor saloon by J. R. (Jack) Johnston, who later was appointed city marshal. The killing was listed as accidental.

About August 1 Newton was indignant over the murder of a young cowboy named Lee who was shot in a dance hall in "Hide Park," the suggestive title given the town's red light district. The youth was a handsome, engaging young fellow with many friends, but the citizens of the town were helpless before the fierce, gun-toting Texans, and this incident, too, was placed on the record as an "accident." About the same time an unidentified man was killed when he was thrown from a wagon while drunk.

Among the hardest individuals who walked the streets of Newton was Mike McCluskie, whose real name was Arthur (or George) Delaney. McCluskie was a night policeman, a fine-looking, squarely built, self-reliant man, chary of speech and bearing a reputation for being tough and ready to shoot.

Wherever McCluskie went he was followed by a thin-faced, sunk-chested youth named Jim Riley. Nobody knew much about Riley, although it was surmised that he was related to McCluskie. He was only eighteen and appeared to be in the last stages of tuberculosis, with a hacking cough and the hectic red in his cheeks of the fever-ridden. McCluskie, who had never befriended anyone else, had been a good angel to this boy. He "grubstaked" him, furnished him a bed, and always took his part in the quarrels that were forever rising. Riley wore old, patched clothing and had no money. He followed McCluskie around like a dog. They made a strangely assorted pair—the fierce, swaggering gunman and the pale, inoffensive youth.

Friday, August 11, was a big day in Newton. An election was being held to vote on the subscribing of $200,000 in county bonds for the building through Newton of the Wichita & Southwestern railroad. William Wilson, alias Bill Bailey, a Texas gambler with two or three notches on his gun handle, served as a special policeman at the election and was thoroughly offensive and officious. During the day McCluskie and Bailey met, and Bailey was drunk. There had been bitterness between them over a woman and now Bailey taunted McCluskie over having been arrested a few days previously on a charge of garroting preferred by Captain French, a Texan. The case was heard before Judge Bowman, who dismissed it for lack of evidence. There was no occasion for the gibe by Bailey, but the thought of the woman over whom they were rivals was fierce between them and they quarreled explosively before they separated.

That evening at 8 o'clock they met again in the Red Front saloon. Bailey, even drunker than before, was very abusive. He demanded that McCluskie set up the drinks. McCluskie refused. Oaths crackled out, followed by blows. Out of the saloon ran Bailey, and crouched across the street before the Blue Front saloon. A moment later Mc-Cluskie stepped out of the Red Front. His six-shooter blazed and the gambler was carried, dying, to a bed in the Santa Fé Hotel, while the girl from "Hide Park," who had been the original cause of the quarrel, sobbed over him.

Newton considered the shooting justified, but the Texas contingent, to which Bailey belonged, thought differently. To avoid trouble, McCluskie left town for a few days to allow feeling to die down. He returned Saturday night, August 19, and in spite of warnings went to Perry Tuttle's dance hall in "Hide Park."

One of the Texas cowboys who felt most bitterly over Bailey's death was young Hugh Anderson, who had just come up the trail with a herd of cattle from his father's ranch. Young Anderson was dangerous and had more than one killing behind him. The night McCluskie returned to Newton, Anderson learned of it and gathering his friends led them straight to Tuttle's place. It was past midnight, the dancing was over in most dance halls and the musicians dismissed, but at Perry Tuttle's the lights still glared brassily.

Within the building glasses clinked, the hoarse voices of the gamblers called their bets, and a spavined piano, a fiddle or two, and perhaps a banjo tinkled out a tune barely discernible above the shuffling feet of the steel-limbed dancers on the floor. McCluskie was taking an active part in all of this, while Riley, with his spasms of coughing, leaned against the wall near the door. Of a sudden the door was violently flung open and Hugh Anderson, with his Texans, stepped grimly in. At one of the gambling

tables sat McCluskie, taking a turn at cards. Straight toward him walked Anderson, his revolver in his hand, looking squarely into the widening eyes of the man he had
come to kill. What he said sounds melodramatic. But the
words were full of deadly realism:

"You cowardly dog! I'm going to blow the top of your
head off!"

The young Texan's voice rose in a sudden furious scream
and the gun in his hand began to cough ear-splittingly. A
bullet ripped through McCluskie's neck and he spun out
of his chair to the floor. A second bullet bored through
his body.

The dance hall was in an uproar, the frenzy of fear.
Cowboys, gamblers, bartenders, and dance girls, who had
filled the floor a moment before, fought to reach the door
or find room behind the bar. In the center of this hysteria
the little group of Texans stood contemptuously, Anderson at their head, his revolver still smoking in his hand
as he looked down at his dying enemy, prostrate on the
floor.

Thus for a moment appeared the scene. Then, without
warning and from a quarter entirely unexpected, disaster
struck the Texas crew.

Jim Riley was leaning near the door by which the
Texans entered. Before he had a chance to move, his friend
and idol, McCluskie, had been shot down before him. For
a moment he looked on, as if he had difficulty in compre-
hending the shocking thing that had occurred. Then the
apparently inoffensive youth went berserk. Jim Riley the
consumptive, Jim Riley the harmless, Jim Riley the butt
of jokes, became in an instant an appalling machine of
death. In two strides he reached the door, slammed it
shut, turned the key. Then he whirled and from somewhere among his nondescript garments pulled a six-shooter.

With his back against the locked door and bitter hate
in his face, Riley began to kill. The reports of his revolver
sounded like the reverberations of cannon in the closed

room. Somebody hurled a chair at the lights, putting the interior in blackness. Still the orange flashes of flame leaped forth from heated gun muzzles.

It lasted only a few moments. Quiet came, as appalling in its way as had been the noise of death. Someone managed to find the door, now left unguarded, and threw it open. A rush of fresh air poured in to disperse the acrid powder smoke. Men and women who had been caught cowering in corners rushed to safety. For long minutes nobody ventured back into the building. Then a bartender had the courage to re-enter and strike a light.

Jim Riley was gone; but on the floor lay the victims of his fury. Nine men were dead or wounded in that room. Beside McCluskie and almost on top of him lay Hugh Anderson, gasping and moaning, it seemed at the edge of eternity. A little to one side sprawled Jim Martin, Anderson's trail boss, dead. Beyond him lay two cowboys, Billy Garrett and Henry Kearns, both shot through the lungs and dying, with Jim Wilkerson, badly wounded, farther on. In another part of the room Pat Lee, a railroad worker, was huddled, shot through the stomach and dying. With him was a track foreman named Hickey, wounded in the jaw and nose, and another bystander named Bartlett with the blood spilling from an ugly hole in his shoulder.

Newton was hysterical with panic and indignation. A doctor was summoned and some of the dance hall girls did womanly service in nursing the wounded. Tom Carson, the city marshal, enlisted a posse to arrest the survivors of the Anderson gang, but the Texans gathered in a threatening knot. It appeared for a time that there would be street fighting as an aftermath of the dance hall battle, but the posse finally dispersed. Even after the withdrawal shootings occurred in different parts of the town. Two white men and a Negro were wounded during the following day, none seriously enough to have their names recorded by the chroniclers of the period.

That Sunday morning, from 8 o'clock to noon, a coroner's jury sat and arrived at a verdict of murder in the case of McCluskie, naming Hugh Anderson as the guilty person. Shortly after the jury adjourned, each juryman received a brief, grim notice:

"Leave Newton at once. If you do not you will be found ornamenting a telegraph pole Monday morning."

The messages undoubtedly were from the Texans, who at this time were talking about burning the town and running out all the prostitutes and gamblers. None of the jurors, however, stopped to question the authority behind the warnings. By Monday morning all were gone, three to Wichita on an early train and the other three elsewhere.

Departure of the coroner's jury removed much of the force of its findings. Hugh Anderson's friends were able to place the wounded Texan on a train where they hid him in a washroom, standing guard at the door all the way to Kansas City. There he received medical treatment and eventually recovered from his wounds sufficiently to travel, but he was permanently crippled and died a few years later.

And Jim Riley? Nobody ever learned what became of him. For a few lurid seconds he occupied the center of a furious drama, a sudden, incandescent machine of death. Then he was gone. He disappeared, where nobody knows, to die in all probability of the mortal illness which had fastened upon him.

The "General Massacre" as it is still called in Newton, was the bloodiest gun battle in the history of the West. Five men died, and seven more, including those shot the following morning, were wounded. It stunned Newton, but the citizens were powerless to remedy the situation.

As the weeks passed there were more shootings. Rowdy Joe Lowe killed Jim Sweet in a street fight. Another revolver duel was fatal to Dan Hicks, with Harry Lovett as the slayer. Tom Carson, the marshal, a nephew of the famous Kit Carson, had served two months as deputy to

Wild Bill Hickok in Abilene, but he now resigned his position. Newton was too tough for him. Captain King was chosen to fill the post. A veteran of the War between the States, the captain was a conscientious officer, but he was "eliminated" November 6 by one Edwards. Then it was that the citizens appointed Jack Johnston marshal.

Johnston was fearless and quick on the draw. Under his injunctions to refrain from gun-toting, there were for months no killings in Newton. But the deadly spirit at length boiled over when a desperado named Pat Fitzpatrick causelessly shot George Halliday, justice of the peace. The murder was in the Gold Room saloon. Halliday had refused to "set up" the drinks at Fitzpatrick's drunken and insulting demand, and the desperado shot him dead. Then turning to the bartender, Fitzpatrick demanded, emphasizing the demand by tapping on the counter with the muzzle of his six-shooter, a big drink of liquor. He received it with electric promptness and there were no quibbles.

Johnston, who had heard of the shooting, strapped on his guns and started for the Gold Room. On the way he met Fitzpatrick coming up the street.

"Halt!" cried the officer. Instead Fitzpatrick went for his gun. He slid forward on his face with a round, blue bullet hole in his forehead as Johnston returned his own smoking weapon to its holster.

That was the last of Newton's bona fide pistol duels, but there was still one more street killing before peace came to the town. Among the better citizens was a young man named Dave Hamill, partner in a general store and one of the organizers of the new Masonic lodge which had just made its appearance in the city. Hamill did not carry weapons but he was not without courage. On Sunday night, February 22, 1873, he was escorting home from church a charming girl, Miss Alice Egy, daughter of a respected citizen. In front of a saloon was a group of loafers and as the couple passed a coarse remark was made.

Hamill turned hotly and expressed his opinion of the man who had made the utterance. A gambler, Jim Shay, stepped forward.

"I made them remarks," he said grimly. "As for you, just make your will, young man. You've got until tomorrow to live."

It should be noted here that Newton was no longer a cow town at this date. The cattle trade had gone and with it the cowboys. Remained behind an offscouring largely from eastern points, who had come to prey on the cowboys and remained stranded in Newton. Shay was of this class.

Unarmed and responsible for the safety of the girl, Hamill walked away. He saw the young woman to her door and hurried, greatly worried, to the office of Cal Johnston, one of his Masonic friends. Never before had he faced death and he wanted advice.

"If he says he's going to get you, there's only one thing to do—get him first," counseled Johnston, a veteran of Sedgwick's corps in the War between the States. A group of Masons was called together.

"We'll see that you get fair play," they told Hamill.

With a Winchester the young man took a stand behind a pile of lumber near the depot. He knew Shay would come that way. As the night hours wore on a group of shadowy figures passed and took up posts on the corners a block away in each direction. The sight gave Hamill a sensation of relief. The shadows were, he knew, his friends forming what has since been known as "the Masonic picket."

Late at night Shay came swaggering down the street. There is little doubt that Hamill took an unfair advantage of him—if to shoot down a man who has causelessly sworn to kill you is taking an advantage. Without a word of warning he opened fire. The first bullet went through Shay's heel. A second shattered his arm. The third was fatal. It punctured his heart. The crowd which came running at the shots was turned back by the Masonic picket. "It's nothing," the curious were told. "Just some harmless shooting."

Hours later Shay's body was found in the street. It was buried in the now populous "Boot Foot Hill" cemetery and with the last shovelful of dirt tossed on the grave Newton's wild days ended forever. On the old Newton court record appears this terse notation:

February 22nd, 1873, Jim Shay, a bad character, and keeper of a dance house, was shot to death by unknown parties.

XVIII

NO SHERIFF WEST OF NEWTON—NO GOD
WEST OF DODGE

In the destiny of the trails many cities took their places, usually for brief periods only, as headquarters of the cattle trade. Wichita succeeded Newton and held the distinction for two years, 1873 and 1874. At the four main highways leading into Wichita were signs:

EVERYTHING GOES IN WICHITA

Leave your revolvers at police
headquarters and get a check.

CARRYING CONCEALED WEAPONS
STRICTLY FORBIDDEN

Where the famous Keno Hall and the New York Store were situated at Main and Douglas was the busiest corner, but a mile west of the town, on the opposite bank of the Arkansas River, stood a lively collection of shanties and tents known as Delano, where the night-life of Wichita really centered. The Newton & Southwestern railroad reached Wichita from Newton in the spring of 1872. Under the supervision of the ubiquitous Joseph G. McCoy, who left Abilene for Newton and left Newton to make his third and last venture in cattle marketing in Wichita, a stockyards was finished by June. At once, because it was nearer than its neighbors to Texas, the town superseded Newton, Ellsworth and Abilene.

As it always was, the cowboys sought relaxation at the end of the trail and this Wichita, with its satellite Delano, was eagerly prepared to furnish. Rowdy Joe Lowe and John (Red) Beard, were the most notable figures in Delano. They ran rival dance halls. Rowdy Joe, who had been in Ellsworth and Newton—killing Jim Sweet in the latter town—was a hard character. Red Beard, on the other hand, was well-educated, a member of a good Illinois family, but slovenly, with a dour, shiftless look, and he carried always with him a double-barreled shotgun.

In Wichita's cattle boom days the dance halls owned by these men were the twin vortices of a deal of lurid action. As early as September, 1872, a Texas cowboy, Jackson Davis, was killed in a pistol fight by Charley Jennison, a Wichitan, who was himself severely wounded. But Delano really began to splash blood in 1873.

There was, for example, the little dispute the night of June 2, when a cavalry soldier from Custer's Seventh Cavalry wanted to dance with a girl at Red Beard's. Unfortunately for this ambition, the girl already was dancing with a Texas cowboy. The cavalryman called his friends to help him and the Texans in the place rallied to their man. Revolvers began to blast in the low-ceilinged room while the girls and other non-combatants stampeded for the exits. When the casualties were counted, one girl was found wounded, one soldier was dead, and two others wounded, in addition to a good deal of glassware and some mirrors smashed. No Texans were hurt.

Whether or not this unfortunate occurrence gave Red

Beard's place a bad name, Rowdy Joe's began to win most of the patronage. Over this circumstance Red Beard began to brood—nor was he the man to permit a thing like this to continue without doing something about it. With a crowd of his friends he invaded Rowdy Joe's, the night of October 27, to break up a dance.

Red's announcement of his program was a shot fired point-blank into the crowd, which wounded Annie Franklin, a dance hall girl. Immediately everyone who could find a gun of any description went into action. The dance hall was filled with flashes of fire and choking clouds of powder smoke while the reports of many guns were almost deafening in the small interior. Rowdy Joe dragged himself to a corner, painfully wounded in the back of his neck. Red Beard fell back and was dragged outside, his arm and hip riddled by buckshot. Bill Anderson who, earlier in the year, had killed a blacksmith in a fight in Wichita, was himself instantly killed by a ball through his head.

In spite of the ebullient outbursts, however, Wichita was able, more than any other cattle town, to meet the lawlessness of the cowboys. A volunteer committee of special police was formed to augment the regular body. E. B. Jewett, a fearless lawyer, was police judge, and W. P. Campbell, was district judge. I knew both of these men well, and Campbell's character is best indicated by the fact that right up to the time of his death at well past ninety, only a few years ago, he was known to everyone as "Tiger Bill."

One of the police officers was Wyatt Earp, who later broke the back of the rustlers' organization in Cochise County, Arizona. It was a potent law-enforcement body. There were some exciting days, but Judge Campbell and Judge Jewett, with the backing of the vigilantes, held grimly and rigorously to their ordained duty. In one summer—that of 1874—Judge Jewett turned over to the city $5,600 he had assessed in fines against cowboys for disturbing the peace, carrying firearms and other things which they considered innocent peccadilloes. Such tactics were discouraging to the trail riders, and Wichita never knew a reign of terror.

As the railroads extended their arms, Caldwell and Hays City took their turns as cattle centers. But at last the real rim-rock cow town came into being. For ten years the cow country knew it as the center of the market, a period longer than that center remained anywhere else—Dodge City, the uncurried, humorously unregenerate, he-coon of cowboy capitals.

Three things made Dodge. First, it was the railhead of the Santa Fé, where the exuberant track crews spent their Saturday night wages and awoke from their debauches in time to go to work again Monday morning, flat broke—the original "hell on wheels," since many of the bagnios, saloons and gambling dives arrived on flat cars and some of them moved on with the steel when the railhead passed on westward. After its start as a roistering railhead, Dodge City became a buffalo hunters' headquarters, from which fierce, bearded adventurers set forth on their pitiless and perilous quest for robes. Finally, Dodge City was a cattle trail terminus. And here Nature came to the assistance of the town. The settlements crowded the trail steadily west, but when at last it swung to Dodge it found itself for a number of years beyond the range of encroachment of the remorseless nester movement. Out that way the plains were high and dry. It was supposed the soil could not be made to grow crops. Simply because the land seemed worthless the frontier stopped short of Dodge and the snaky cattle tracks, winding up from Texas, found there a market for the herds which came north over them.

It is a matter of history that the first cattle herd from Texas passed over the site of Dodge City in 1872, three years before the market was opened there. D. W. Barton, known to his friends as Doc, who, as this is written, still lives at Ingalls, Kansas, brought that first herd north. He started his drive near Pecos, Texas, in March, 1872, and followed the Goodnight-Loving trail as far as Pueblo, Colorado. His first intention was to sell in Colorado, but there was no market for cattle there, so he turned east along the right bank of the Arkansas, crossed over to the left bank about the middle of August, near Granada, Colorado, wintered the stock under the bluffs to the north of the river near the present site of Pierceville, Kansas, and the following spring drove through Dodge City. At that time a railroad ran through the town but there were no stockyards, so Doc Barton took eight hundred of his cattle as far east as Great Bend, where he disposed of them in 1874.

The balance of his herd he kept and became one of the big cattlemen of the upper Arkansas valley. In 1877 he returned to Texas and brought back to Dodge not only three thousand additional cattle, which included his wife's dower herd, but Mrs. Barton herself. Mrs. Barton drove the chuckwagon every step of the way—a feat considerably more difficult than guiding an eight-cylinder sedan down a concrete pavement. It is interesting that the trail laid out by the Bartons across the Indian Territory became one of the main branches of the Western trail, and on the basis of it Doc Barton was chosen chairman of the trails committee of the Cattleman's Association of Dodge City, spending much of his time the next few years laying out other routes from Texas.

The cattle market opened at Dodge when the stockyards were completed and the first important drive was in 1875. Because it was the greatest and longest-lived cow town, Dodge deservedly has the greatest reputation. Nobody knows how many persons, of both sexes, died by bullets in its streets, saloons and resorts. Robert L. Wright, an early mayor of the city, in his book *Dodge City, the Cowboy Capital*, said there were twenty-five killings in the first year of the cattle drive. Bill Brooks, who was for a few weeks marshal of Newton and later went to enforce the law in Dodge City, is credited by Floyd B. Streeter, in his *Prairie Trails and Cow Towns*, with shooting fifteen men the first thirty days he was on duty. Probably not all these were killed. Everett Dick, who made a scholarly study of the subject, sets the more conservative total of fourteen as the number of gun-shooting deaths the first year.

Dodge City's celebrity in the West is shown by some of the generic names it gave the world. The first "red light district" was in Dodge. Until a few years ago the old Red Light House still stood south of the tracks, although long unoccupied. It was a two-story frame structure with a blood-red glass in the front door through which the light shone in lurid welcome to the celebrating cowboys who ranged the streets outside. Other similar establishments existed in the vicinity and the entire area was called the Red Light district from this one house. The name has since gone all over the world. There are red light districts in far Korea, Singapore, Paris, Buenos Aires, and exotic ports throughout the seven seas today.

A second name Dodge City gave the world was Boot Hill. Today Boot Hills are to be found all over the West. Even Newton has a "Boot Foot Hill" which antedates the Dodge City cemetery, but the name was given it after Dodge City's burying ground became famous. First to occupy Boot Hill, which lay on a knob overlooking the lower part of town, was a Negro called Texas, who was shot by a gambler with the equally noncommittal name of Denver. Last to be buried there was a dance hall girl named Alice Chambers, to whom the town gave a funeral on May 5, 1878.

No Western town ever had the array of gunmen as peace officers that Dodge City possessed. Wild Bill Hickok, contrary to many accounts, never was marshal or even deputy in Dodge. He was in the town, but as a private citizen, making his living by gambling. But almost every other noted six-shooter expert in the country, at one time or another, wore the star in Dodge City. A mere listing of their names sounds like a *Who's Who* directory of plains fighting men: Ed and Bat Masterson, Mysterious Dave Mather, Bill Tilghman, Pat Sughrue, Bill Brooks, Ben Daniels, Wyatt Earp, Jack Allen, Charlie Basset, and Tom C. Nixon were a few.

First, and one of the bravest, was Ed Masterson. When he took office the saying was that Dodge "had a man for breakfast" every morning. The marshal tried to put an end to the slaughter. Twice he was wounded in gun fights with desperados, dying the second time.

Bob Shaw and Texas Dick Moore were having an ugly altercation in the Lone Star dance hall one day in November, 1877. When Masterson arrived he found Shaw "lickered up" and nursing a great resentment because he had lost forty dollars, gambling with Moore, whom he was covering with his drawn revolver. The marshal ordered the cowboy to turn over his weapon. When the order was ignored, Masterson "pistol whipped" Shaw.

This was a most ungentle action. Western gunmen are sometimes represented by the motion pictures as clubbing their victims by holding the barrels of their weapons and striking with the butt. No real Westerner would have been guilty of doing that. It would have rendered him unable to shoot; it would have been dangerous to himself if the gun happened to be discharged by the concussion of the blow; and the very act of changing from a normal grip to a grip on the barrel requires time, and time was a thing which was very precious under the circumstances. "Pistol whipping," also called "buffaloing," was done with the revolver held in the ordinary manner, the long heavy barrel being brought down with a whipping motion on the skull of the antagonist.

Usually it was very effective, as when Wyatt Earp pistol whipped a whole gang of gun-fighters into submission in Tombstone, but on the occasion in the Lone Star dance hall Masterson did not strike hard enough. Bob Shaw remained on his feet and, whirling, shot the marshal through the body, paralyzing temporarily his right arm. As Masterson fell, he coolly transferred his six-shooter from his useless right hand and, firing with his left, punctured Shaw's left arm and left leg, Texas Dick in the groin, and Frank Buskirk, an onlooker, who had not found cover quickly enough, in the left arm. All the wounded recovered, but the incident illustrates the coolness and courage of the officer.

The marshal's second battle, April 8, 1878, did not end so fortunately. Jack Wagoner and Alfred Walker, on a spree, were terrorizing the town and Masterson attempted to pacify them. Drunk and deadly, the cowboys killed the marshal. But he did not remain long unavenged. His brother Bat Masterson, arriving in some haste as the killing occurred, shot just twice. Wagoner and Walker dropped dead. The three men were killed in the space of a few seconds. Bat Masterson was one of the great six-gun experts and a member of the buffalo hunter crew which, in 1874, defeated the allied hostile Indian tribes at Adobe Walls, Texas.

Another old buffalo hunter was Kirk Jordan, who usually carried about with him his heavy buffalo gun with the huge .50 caliber bore. Nobody could scare Jordan—not even Bill Brooks, whose fearsome reputation as a killer has been noted. When the two quarreled the old buffalo hunter went for the gunman as if he were a tenderfoot. Down the street Jordan chased Brooks, who ducked behind one of the barrels, which, filled with water, stood on Dodge City's street corners for use in case of a fire. Jordan fired and the heavy Sharp's bullet ripped right through the barrel, water and all, but was so spent that it did not harm Brooks.

It did, however, chill his blood. Again the gunman took to his heels. He reached Bob Wright's livery stable. Wright hid him under a bed until night, when he was taken to Fort Dodge, from whence he left the country for good. William

McLeod Raine quotes a report that Brooks was lynched a year or so later by vigilantes in another Western town.

Two of the most deadly gunmen in Dodge were Tom C. Nixon and Mysterious Dave Mather, and each served a term as peace officer. There was considerable divergence of opinion among the citizens as to which of the men was the quicker and surer with a six-shooter. The question created much interested speculation, with perhaps a few friendly wagers, contingent on the matter ever coming to a satisfactory adjudication. Speculation ended when, on July 21, 1884, the two quarreled over a woman. Everyone considered that Mysterious Dave furnished his rival with a very decent funeral.

Early in 1879 the Dodge City *Globe* published this statement:

Colonel Straughn [the coroner] has removed thirty-three bodies from the [Boot] hill. He found all of them in a good state of preservation. Their boots had been removed and placed under their heads for pillows.

Those thirty-three bodies were not all that occupied Boot Hill. In later excavations other skeletons were found, and the nearest total which can be arrived at is forty. Most of them were persons who died violently, "with their boots on," as the Western saying was. Nor was this the only place where gun battle victims were buried. I have a letter from Merritt Beeson of Dodge City, possibly the best of all authorities on this particular subject of morbid frontier interest, which says in part:

From 1872 to 1878 there was no regular burial ground in Dodge. Important persons dying in Dodge, who had friends, relatives and money especially, were taken to Fort Dodge, four and one-half miles down the river, and buried there. Others who were of no consequence were stripped of valuables and wearing apparel worth saving, then rolled in blankets and buried anywhere about town, not Boot Hill in particular—just a soft place to dig.

Not all the occupants of Boot Hill were men. Two at least were women. Alice Chambers has been mentioned. The other was Lizzie Palmer, also a dance hall girl. Mrs. Annie Anderson, who died in 1931, and was in her own youth a dance hall girl in Ellsworth, Newton and Dodge, as the trails moved west, recalled that Lizzie Palmer died as a result of a fight with another girl. There was some scratching and hair-pulling, and Lizzie received a scalp wound which became infected. She died from blood poisoning. A big crowd of cowboys was celebrating in town. When the news of Lizzie's death was announced, the men, always sentimental, voted to give her a big funeral.

"The Methodist minister refused to conduct the funeral service," said Mrs. Anderson, "but another preacher who was in town for his health agreed to do so, saying the worse they were, the more they needed the rites of the church. So they buried her on Boot Hill. The cowboys furnished that preacher with all the money he needed to stay West for several years after that. When he returned home he was given a big escort to the train. Cowboys liked to do things up right."

The most celebrated Dodge City tragedy involving a woman was the killing of Dora Hand. Her real name was Fannie Keenan, but Dodge knew her as Dora Hand and she remains today the town's greatest figure of glamour and mystery. Her previous history was known only to one man, James H. (Dog) Kelley, the first mayor of Dodge, and owner of the Alhambra saloon and dance hall where she entertained. By all accounts she was a beautiful creature, with a face and voice which gave men strange nostalgic dreams of better days and finer surroundings, and her charm was such that twelve men are said to have died for her smiles. That charm was fatal to her in the end.

Old-timers have said that Dora Hand was reared in a cultured Boston home, educated in music in Germany, and that she had had success on the stage before audiences far more sophisticated than any in Dodge. But there was a strange, wild streak in her which took her pretty little feet down a path that eventually led to death. Dora had sung on the Comique stage with Eddie Foy. Before she went to Dodge she had been at Abilene and Hays. She was a part of the seamy, sordid life of the border, yet no woman in town was more willing to sing at funerals, weddings or other occasions than she.

The last victim of her bright face was Jim Kenedy, commonly called Spike. There are varying tales of what occurred. One is that Dog Kelley, resenting Kenedy's too-ardent attentions to his charming protégée, caused the young cowman to be ejected bodily from the Alhambra one night. For that Kenedy, a Texan, swore to kill the saloon man.

Dora was living in a house belonging to Dog Kelley—a flimsy, two-room shack, the front room of which he rented to his star entertainer and the rear one to another of his girls, Fannie Garretson. Early the morning of October 8, 1878, the Texan crept up to Kelley's house and fired two shots at it, aiming where he thought Kelley would be sleeping. One bullet lodged in the floor. The second passed over the bed occupied by Miss Garretson, cut through the partition, and struck Dora Hand. She died instantly, a dainty mystery to the last.

Her funeral was the largest ever held in frontier Dodge. When the wagon containing the coffin—the best the town afforded—moved up from Front Street toward the new Prairie Grove cemetery northwest of town, it was followed by hundreds—dance hall girls, cowboys, gamblers, gunmen, business men, wealthy ranch owners, "respectable" women, professional men—the elite as well as the off-scourings of the town. The only minister in Dodge at the time took for the text of his sermon, "He that is without sin among you, let him first cast a stone at her."

Meantime Kenedy was pursued for a hundred miles by Wyatt Earp and Bat Masterson and captured at last after his horse was killed. But he had powerful connections in Texas, and he was eventually saved by the money of his family and permitted to return to Texas. When Dog Kelley died at the Fort Dodge old soldiers' home in 1912, he still refused to tell the story of the early life of Dora Hand. He carried her secret with him to the grave.

So the population of Boot Hill and other burial places in Dodge City grew, largely because of differences between gentlemen quick on the draw, as the great herds of cattle, like rivers tossing upon their surfaces a flotsam of sharp horns, poured slowly north, bringing strangers to town. The incoming population received its wages when Dodge City was reached and the gamut of wild entertainment of a frontier town awaited them. Frequently a long session of debauchery ended with the unceremonious dumping of a body in Tin Can Alley, which is now Chestnut Street, where it would wait its turn to be carted off to the burial ground.

At first the unwritten law of "first on the trigger" met little check. The West, and particularly Dodge, looked with some lenience on a custom that relieved the peace officers of not a little responsibility. When two men settled a question with guns, the incident usually was pretty definitely closed. An officer had to be careful in those days, like everyone else. If a decision was made contrary to the judgment of public opinion, there was little to prevent a marshal or justice of the peace from being "recalled" by the lethal end of a six-shooter.

But where the deliberations of Dodge City jurisprudence failed to bring peace, Mother Nature in one sweep did so. That was the day, early in January, 1886, when the great blizzard swept out of the Northwest. Into every nook and crevice the snow sifted, and for days the king of storms lashed the plains. When it was gone the cattle—all except a few stragglers which survived by some miracle—were dead in the snow. The kingdom of the cowmen in the South-west had toppled. Dodge City was the capital of a broken empire, with little to wait for except the slow coming of the sod-buster plow.

4. . . THE COWMAN TAKES OVER

XIX

PUTTING DOWN STAKES

Major Andrew Drumm, John W. Iliff, and Charles Goodnight were three kings in any deck. They possessed similar qualities of imagination, foresight, judgment, and that indescribable yearning for danger as a thing to be savored for itself alone, which taken together make the great adventurers. As they were the leaders, they were also the patterns for the three great waves of cattle range expansion which took place in the '70's and '80's.

Trail herds by the hundreds had poured north across the Indian Territory between 1866 and 1870, but the latter year was the first in which Andy Drumm brought up cattle from Texas. He was an ex-California gold miner, hailing originally from Zanesville, Ohio. Gold he had found, built a snug fortune, then turned to livestock on the Pacific coast. By 1869 he had helped organize one of San Francisco's pioneer meat-packing companies, Willoughby Bros. & Drumm.

Above everything, however, Andy Drumm craved action, and after a time he wearied of sure and steady business and began looking around for adventure. Trail driving was then in full blast in Texas. It combined two things Drumm wanted—a chance to make money for the man who was willing to gamble, and excitement. He crossed the mountains and deserts from California to Texas in 1870.

"We used to trail cattle up from Texas and double our money on them," he used to say in his later days in Kansas

City after he became a multi-millionaire in the livestock
commission business. "We didn't mind the distance. The
range was free and there were no fences, so feed cost us
little."

To hear him tell it, it was as simple as that. Not a word
of "big swimming," of sudden stampedes in the night when
men were trampled to death under frantic hoofs, of Indian
scares and bandit raids, of cattle thieves and granger opposition, of six-shooters flaring in frontier saloons, and the
other manifold risks of the trail, both physical and financial. "We didn't mind the distance," was his word.

As stated, 1870 was Drumm's first year up the trail. The
first night he bedded his herd in the Cherokee Strip he
took note of the marvelous possibilities of the country for
cattle raising. He observed its luxuriant grass, its water, and
its mild climate. Other trail drivers had noticed these things
also—they were too obvious to be overlooked by experienced cattlemen. But with Drumm, to understand possibilities was to wish to profit by them. He had a passion for
immediate action. Before the year ended he had stopped a
herd in the Cherokee Strip and settled there. On that day
Andy Drumm became the first cattle grower to ranch in
that famous area.

The Cherokee Strip was a curious geographical oddity.
It stretched sixty miles wide by two hundred and fifty
long, from east to west across the northern part of Indian
Territory. Originally it was laid out by the government
as a corridor to permit Cherokee Indians to travel west
to the buffalo country across their own land, thus avoid-
ing clashes with the troublesome and predatory Kiowas,
Comanches and Cheyennes. But the lazy Cherokees cared
little about buffalo hunting after the novelty wore off, and
as punishment for their aid to the Confederate forces during the War between the States, other tribes were settled
west of them so that they were cut off from their own
outlet. By the time Drumm arrived in the country it was a
huge expanse of uninhabited territory, lying idle, swarm-

ing with game, sometimes fiercely burned over by prairie fires—simply begging for someone to make use of it.

This Andrew Drumm did. He was quickly followed by Colonel Evans of St. Louis, E. M. Hewins of Cedarvale, Kansas, and a number of others who soon occupied the entire Strip with large ranches. Drumm and his partner, Andy Snyder, ran the U brand on about 150,000 acres, and sometimes held 50,000 head of cattle on it, although of course that could be done only for short periods. Intersected by the Salt Fork and Medicine Rivers from the north and Cottonwood Creek from the south, the location was ideal, with timber along the streams to afford shade in summer and protection in winter, and the peerless advantage not afforded by most large ranges that it was only a few miles to water from any spot on it.

Among other things, one made Drumm notable: he is believed to have been the first big rancher to put up barbed wire extensively. Shortly after the "Glidden wire," as it was called, was patented, the major built a drift fence to the north of his range. Later he fenced in his entire acreage and thereby started something which eventually spelled doom for the whole free, wild way of life he represented. But that is getting ahead of the story. Drumm set a great cattle trend by ranching in the Cherokee Strip. It influenced much of the future spread of cattle through Western Kansas and Oklahoma.

In a similar manner, Charles Goodnight foresaw the possibilities to the west of Texas. When he and Oliver Loving first plunged across the terrible Horsehead trail in 1866, they established a wave of cattle migration. Before long Goodnight changed from trailing to raising cattle and contracted with John Chisum to bring herds across to New Mexico. Presently Chisum followed into the Pecos valley, and other ranchers arrived to establish a vast cattle domain there.

After the death of his partner, Goodnight moved up to

Colorado where he established two ranches. Grasses on the slopes of the Rockies added generously to the tallow on a steer, as he discovered. In 1868 he purchased a large acreage not far from Pueblo and in 1869 took another property forty miles from Trinidad. His entire life's earnings, $72,000, went into these ranches and the cattle to stock them.

A panic and a blizzard ruined Goodnight, but he had established a mighty westward trend of cattle movement out of Texas, which extended throughout Arizona and New Mexico as well as Colorado and northward. There is, moreover, a pleasant and intensely human story of how he recouped his fortunes and opened still another cattle empire in later years, which will be dealt with in another chapter.

Even preceding Goodnight and Drumm was John W. Iliff, who had been a student in the ivied halls of Delaware College until, in 1859, he struck west. That year he reached Colorado Territory with a wagon-load of provisions which he sold for $100. Iliff had an idea. By no means was he an experienced cattleman, but he was better than that—he was a born cattleman. He saw about him a potential range country and his hunch was as good as the judgment of years.

He established a store on the Oregon trail and began purchasing a little herd of cattle—off-throwings from the emigrant trains. On the great transcontinental thoroughfares livestock often became footsore, sick, or worn out, and could travel no farther. Many a trading post was established along the trails which made a good profit in buying worn or ailing cattle and horses from emigrants at panic prices, selling in exchange sound animals at boom prices, and permitting the unsound stock to recuperate, when, in a few weeks or months, they again were sold at outrageous prices to newly arrived emigrants who in turn left behind their debilitated animals.

Iliff did not go into this sort of thing. His idea was to raise cattle as a business in itself, and sell them to the markets. Along the South Platte he ranged his slowly growing herds, and when the claim laws took effect he began a gigantic project to control all the water he could grab. For thirty-five miles along the Platte he had claims taken up. On some of it he filed himself; his cowboys, friends, perhaps even a few fictitious individuals, filed on the rest. In the end Iliff controlled thirty-five miles of riparian rights, and that gave him control of the land as far back as the cattle would walk to graze and return to water. It was a superb example of seizing water rights wholesale, the secret of many an early cattle fortune.

Iliff purchased many cattle from Goodnight and Loving. The first herd of stockers the latter took up the Arkansas valley was bought by local farmers, the demand being brisk owing to discovery of gold in the Pikes Peak district. Loving's second stocker herd, after the first Horsehead drive, was purchased in part by Iliff, and when Goodnight, after his partner's death, continued the northern drives, the cattle went to Iliff who took them on the Platte.

Before he was somewhat constricted by the title laws, Iliff is said to have ranged for a hundred miles up and down the South Platte, with Julesburg as a center, his market for beef being Denver, the railroad construction gangs, the Indian reservations, and the mining country. He was one cattle king who attended personally to his business. While other ranchers took ocean voyages to Europe or lolled at the Cheyenne Club or in Denver or Kansas City, John Iliff day after day doggedly rode the rounds of his ranch properties. One of his boasts was that he could ride his range for a week and sleep in a different headquarters building each night. He was never known to carry a gun. His word was law, and his jealously loyal cowpunchers enforced it.

It was natural that Iliff, like Drumm and Goodnight, should be followed in his idea. The greatest of all ranges

—Wyoming, Montana and the Dakotas—was still to be stocked. Of an estimated ten million longhorns that went up the great trails from Texas in the twenty-three year period from 1867 to 1890, by far the larger part was destined to become stockers on the range.

Drumm, Goodnight, Iliff. They were symbols of a magnificent expansion; leaders in a migration which remade the map and history of the continent. Yet that remaking awaited the removal of two great deterrents. Even with the tides of cattle already beginning to surge, there could be no full flood of dust-raising, bellowing, horn-clicking herds loosed on the North and the West until the Indian and the buffalo were gone. In the necessary abolition of those twin barriers to civilization was tragedy and cruelty. The destruction of the bison broke the spirit of the red people who depended on the wild herds for their commissary. Neither the extermination of the buffalo nor the crushing of the Indians was a cattleman's accomplishment, although the riders of the range played a part in each. But both these great acts in the drama of the West were intimately bound up with the fate of the cattle country.

XX

LOS COMANCHEROS

The word *Comancheros* has almost no recognition in formal history, yet it is a word once excessively well known to the unprotected ranchers on the scantily settled borders in Texas. Comancheros and Comanches became almost synonymous in import and significance, although the two peoples were of different race. To the door of the Comanchero can be laid part—by no means all—of the tragic harm the Comanche wrought.

Before the Texas longhorns had even approached the escarpment of the *Llanos Estacados*—a vast step in the level land which runs north and south across most of Western Texas at about the one hundredth meridian—and before white men from the East or South had made any gesture toward exploring the high plains above the escarpment, Mexicans from the west were amazingly familiar with the strange flat sweep of the Staked Plains, with the surprising gashes of canyons which score and furrow across that area. First to come to know the *Llanos* were Mexican buffalo hunters from Santa Fé, Taos and other New Mexican settlements. They called themselves *ciboleros* and hunted, after the manner of the Comanches themselves, the bison with the lance. The *ciboleros* butchered and dried the meat they killed, packed it in rawhide sacks, and carried it to the settlements, where as *carne seco* (jerked beef) it was an important article of commerce.

Gradually, in this manner, trade relations were established with the Comanches, who became accustomed to the visits of the Mexicans to the buffalo country and learned to exchange peltries and other articles with the visitors. After a time a new direction was given Comanche war activities by the type of trade in which the Mexicans displayed interest. Buffalo hunting was all but forgotten. The men who left the Spanish settlements for the Staked Plains became known as Comancheros—those having to do with Comanches. Because they furnished an avid market for stolen cattle, horses, and other plunder, they were a tremendous spur to the Indians in raiding the Texas frontier.

As early as 1832 Zebulon Pike reported seeing the deep trails made by the Comanchero *carretas* across the plains toward the Comanche country—carrying out goods to be exchanged for loot. Notice was taken of the trails, campsites and the slinking traders themselves by Josiah Gregg in 1839, and Captain R. B. Marcy in 1849. At first the Comancheros made up their expeditions and went out into the *Llanos* with no definite destination, hoping to run into some wandering band of Comanches. But by the time of the War between the States, the trade had become so mutually satisfactory that regular rendezvous were appointed where the Mexicans with their goods and the Indians with their plunder met by agreement and exchanged commodities. Names of some of those rendezvous still stick to localities in the plains and they are revealing. One is Las Lenguas, or Tongue River. Here often gathered a hodgepodge of Indian tribes—Comanches, Kiowas, Lipans, and the like—with Mexicans and even renegade whites, in a native American Babel of the plains whose varied jargons gave the place the name River of Tongues. Near Quitaque is a valley called Valle de Las Lagrimas—Valley of Tears. Here the Comanches, after their kidnaping raids, during a period when abduction of white women and children became a veritable industry for the ransoms to be obtained, tore weeping mothers and children apart as the captives were apportioned to different bands which scattered to various widely separated localities. The tragedy of these unfortunates left its impression even on the callous Mexicans who witnessed it—hence the name.

When the grass grew green in the spring the Comanches would ride down on the Texas cattle country, sweep away with horses or cows, drive the stolen herds north and west, and meet the Comancheros at the places mentioned, or other points equally well known to them. There the bartering would take place. At times the Comanches cynically offered to the Mexican traders the persons of Mexican captives, particularly women, whom they had taken in raids down in Mexico. These the Comancheros equally cynically refused, unless they had use for them. There was about them no silly sentimentality concerning their own people such as caused white men to pay big ransoms or set armies in motion to rescue a white child or an enslaved woman.

More often, however, the loot to be sold was cattle or horses. Live Texas cattle on the hoof were more profitable and easier to handle than jerked buffalo meat in sacks. Brisk also was the trade in horses and mules. Shortly after the close of the War between the States Charles Goodnight obtained from some New Mexicans the admission that not less than three hundred thousand cattle and a hundred thousand horses and mules had been bartered from the Indians by them. That stock came from Texas.

Sometimes unscrupulous Americans financed the New Mexican traders. So well established was the traffic that occasionally persons participated in it without realizing its implications, as when the wife of an officer at Fort Bascom sent a copper kettle and a Navajo blanket with Mexican traders in 1865, perfectly unaware of wrongdoing, and in return found herself possessed of twelve head of cattle.

The incident emphasizes a point. However they obtained their loot, the Comanches nearly always were woefully cheated in disposing of it to the Comancheros. A keg of whiskey bought a prime mule, ten pounds of coffee went for a pack horse, a buffalo robe brought little or nothing, cattle what they would fetch. Muskets, powder and ball, and knives were sought after by the Indians as eagerly as they sought whiskey.

After Oliver Loving's death in 1867, Goodnight found on Gallinas Creek, in New Mexico, six hundred of the cattle stolen from him previously at the Horsehead crossing. They were in the possession of Mexicans who had traded for them with the Comanches, and although he went to court to replevin his property he not only lost his case in Las Vegas, but had to pay the court costs of $75, and was happy to get off as cheaply as that.

There was one brief period of retribution when, in 1872, John Hittson and H. M. Childress, with sixty hard-faced Texas cowmen, armed with powers of attorney from many ranchers in the Lone Star State, rode into New Mexico to look over the cattle there. Goodnight refused to join this extra-legal expedition, but he probably gave it his blessing. Individualistic and defiant, Hittson and Childress swept through that country repossessing cattle wherever good ownership credentials could not be shown, and rode back at last with eleven thousand head which they had recovered. Even this, however, failed to halt the trade across the Staked Plains.

Thus in fancy-free and frisky emancipation the Comanches and their numerous copper-colored allies disported themselves on the Southern plains, with a pleasant source of plunder and a goal for raiding on the one side, a ready market for surplus booty on the other, and a steady supply of food on the hoof in the buffalo herds ready to hand.

And in an even larger area to the north, the Sioux, Cheyennes, and their allies, also viewed life as pleasant, with
ample opportunity for healthful and exciting horse-stealing
from the whites, the occasional prospect of picking up a
scalp, warm canyons into which to withdraw in winter—
and again the ever-present buffalo to furnish robes, meat,
and many other necessary things.

The buffalo was the key to the Indian problem as was
stated in so many words by General Phil Sheridan, and that
problem seemed enough to baffle the wisest. Nobody knows
how many bison were on the plains at the peak of their
numbers. Ernest Thompson Seton, the naturalist, has estimated seventy-five million ranging from Northern Mexico
to Central Canada and from the Alleghenies to the Cas-
cades, in the period just before the Indian acquired the
horse. In the '60's General Sheridan computed that there
were one hundred million in the Southern plains alone. Yet
in a space of one decade those vast herds were wiped out,
and in a manner so cold-blooded and methodical that it left
the world aghast. The buffalo were slaughtered, not by the
soldiers who wanted them out of the way to make easier
the defeat of the Indians; not by cattlemen who wanted
them out of the way to make room for cattle range; but by
a new and potent army—an army of bearded, uncouth
men, who carried great rifles with huge bores—buffalo
guns—and who were, to use the phrase of their legitimate
precursors, the Mountain Men, "half hoss, half alligator."

No exact parallel to the buffalo hunter exists. Many of
them were of high character and fine personality like Billy
Dixon, J. Wright Mooar, Pat Garrett and others. But many
also were murderers, thieves, fugitives from justice, who
sought for good reason the anonymity of this wildest of
professions. One of the hunters killed by the Indians near
the Adobe Walls, in 1874, was an heir of a British noble
house whose real story still remains a mystery, but whose
death created serious diplomatic complications.

Always the Indians had conducted buffalo hunting on a

large scale, and so did meat-hunting whites like William F. Cody, the famous Buffalo Bill. But it was not until the sudden discovery in 1870 that the hides were commercially usable that the real slaughter began. J. Wright Mooar, who at the time this is written still lives near Snyder, Texas, is credited with bringing the buffalo hide before the world as an article of commerce and thus creating the market which was the one thing needed to bring about the destruction of the herds. Mooar it was who sent a shipment of hides experimentally to a New York tanning firm, which found them usable, particularly for machine belting.

Out on the plains in the '70's went the buffalo hunters by the thousands. A good hunter could kill a hundred and fifty to two hundred animals a day—the number limited only by the ability of the skinners to keep up with him. So many hunters were at work all the time that old-timers have told me that on a clear morning the firing sounded like a fair-sized battle in progress. Following the hunter were his skinners, working in pairs, each pair with a team of horses or mules. A knife split the skin down the belly and legs of the dead buffalo, and the team, hitched to an edge of the hide, peeled it off. Pegged down and dried, it was eventually baled with other hides and taken to shipping points.

The Indian Territory and the Texas Panhandle were penetrated by hide hunters in direct violation of the Medicine Lodge treaty of 1867, in which the United States pledged itself to keep its citizens from hunting south of the Kansas line. So the Comanches and their allies went on the war path. It was a disastrous mistake for them. Times had changed vastly since the early days of Texas. The Indians killed a few isolated settlers and freighters that summer of 1874, but otherwise they did not win a single victory against the white man. Their first and most promising chance for a triumph was nullified by a curious coincidence when the ridge-pole of a sod-house saloon cracked, waking the people in the buffalo hunters' fort at Adobe

Walls the morning of June 28, 1874—just before the Indians swept down on the place. The twenty-eight men and one woman in the fort might have been surprised and killed but for this accident. As it was the buffalo guns roared out an immense thunder of slaughter and the tribes were beaten off, carrying their bleeding dead.

Elsewhere Colonel Nelson A. Miles, Colonel Ranald S. MacKenzie, Lieutenant Colonel John W. Davidson, and other army commanders marched and counter-marched, back and forth with their forces through the Indian country, finding and chastising the hostile tribes. Dazed and shocked by the turn things had taken, the Indians doubled and hid, but they were hunted down relentlessly.

It is a pleasure to record that on one of MacKenzie's marches a Comanchero had a most unpleasant experience. MacKenzie, a dour and direct Scot, encountered old and wealthy José Pieda Tafoya, one of the chief Comancheros, riding comfortably across the plains with his well-laden caravan, expecting to meet his Comanche friends and make some profitable transactions in gewgaws for the blood-spattered plunder of the Southwestern frontier. But MacKenzie immediately put Tafoya under arrest and began to question him on the whereabouts of the Indians. With dignity Tafoya refused to talk. The colonel, who possessed a sense of elemental justice, caused the tongue of an army wagon to be propped high and from it swung Tafoya by the neck. Not as high as Haman was the old Comanchero, but he was fully as insecure in his footing and as short of breath. In a single cerebral process he both discovered the use of his tongue and forgot his dignity. When he was lowered to the ground he was voluble with information, and through him the army found the main Comanche village in the Palo Duro canyon and scored the most decisive blow of the campaign.

By the end of the strenuous little war of 1874-75, the Indian question was pretty well settled on the Southern plains. This end was achieved in part by the fact that most of the buffalo were gone. Some were left, but even these were practically exterminated in the Southern grazing grounds by 1880, and thereafter the Comanches and other tribes were happy to abide peacefully near an agency where they could be sure of eating regularly. At last the cattle herds could begin to move.

A very similar series of events was in process in the Northwest. There it was the Sioux who were the chief bar to the opening of the great grazing country of Wyoming, Montana and the Dakotas. Red Cloud and his warriors bloodily forced the abandonment of the Bozeman trail, the route followed by Nelson Story in his great trek of 1866, and no more cattle thenceforth moved up it for a long time. After he saw the government forts burned when the United States abandoned them as a condition of peace with the Sioux, Red Cloud retired on his laurels and was succeeded by a group of even more implacable foes of the paleface—Sitting Bull, Crazy Horse, Gall, Little Wolf and their wild fellows.

Of those barbaric paladins Sitting Bull was the most famous, but Crazy Horse probably was the better fighter. At least the latter is credited by most authorities with furnishing whatever guiding genius there was in the battle on the Rosebud River, June 17, 1876, where Colonel George Crook had his bellyful of fighting painted warriors, and the battle on the Little Bighorn River, June 26, 1876, where Lieutenant Colonel George Armstrong Custer and his detachment of the Seventh Cavalry were cut off and killed to a man.

This last was the greatest victory ever won by Indians west of the Mississippi over white men. Two hundred and sixty-five men of Custer's command were killed. But it was a Pyrrhic victory, for from that date the power of the Sioux steadily declined. A vast deal of marching, scouting and occasional fighting remained to be done. The Cheyennes under Little Wolf were crushed by MacKenzie in

Crazy Woman Canyon. Miles first chased Sitting Bull up across the Canadian border, then defeated Crazy Horse at Wolf Mountain. The end was predestined from the day of the Little Bighorn. Gunpowder and lead had the disadvantage of not growing in the Sioux country as did the willow switches from which arrows were fashioned, and when the Indians used up their ammunition in that climactic battle, they never again were able to gather together enough powder and ball to fight a major action. Wherefore they surrendered.

Other tribes learned the same bitter lesson—that they no longer could contend with the white army. The Piegans and Blackfeet learned it from Colonel E. M. Baker in 1870, and the Nez Percés learned it from the indefatigable Colonel Miles in 1877. With the surrender of the last bands of Sioux in 1878 the Northwest was open for the cattle industry—except for the buffalo which still blackened the Northern plains.

Not until 1880, when the Northern Pacific railroad extended its line from Bismarck west through the buffalo country, did the destruction of the northern herds begin in earnest. When the transportation at last was provided, however, more than five thousand hunters and skinners flooded out on the plains by 1882, especially after it was learned that the hides of northern buffalo were bringing at the market two or three times as much as those of the southern animals. By 1883 the extermination was virtually complete. So low had the kill fallen that year, that in spite of a continuing and greedy market, only forty thousand hides were shipped east. In 1884 two miserable carloads, containing about three hundred skins, were sent from Dickinson, North Dakota—the last wholesale shipment of buffalo hides.

On the plains remained only great racks of whitening bones to remind the beholder of the noble herds that once had covered the high steppes of America. Millions of tons of these bones were destined later to be picked up by "bone pickers" and sold to fertilizer companies. By 1889, fifteen years after hide hunting ceased for lack of sufficient animals to hunt, Dr. William T. Hornaday in a world-wide survey, discovered that only 1,091 American buffalo were in existence out of the seventy-five million which once had grazed in interior America. His book, *Extermination of the American Bison*, brought the first awakening of the public. Fortunately it came in time. The American Bison Society was formed; numerous national leaders, including President Theodore Roosevelt, were enlisted; and today the bison is secure from extermination, although it is his fate for all time to occupy game preserves and parks.

The cattlemen did not destroy the buffalo, nor was it at his instance that the destruction occurred. But the result was as if the cattleman had planned and carried it out. Spotted herds of longhorns were already moving into the buffalo lands even before the last of the shaggy, hump-backed horde had vacated them.

XXI

LEAN YEARS AND LEAN KINE

To return to the cattle country: the year 1873 was one of disaster. It began with lofty prospects after the good year of 1872, in which, besides favorable prices and a brisk demand for beef, new and closer shipping points to Texas were established. Because of the optimism engendered by 1872, more than half a million cattle took the northern trail from Texas the following year, their owners and cowboy outfits buoyant with anticipations of a favorable season.

But it was one of those wet, rainy springs, when every river was "big swimming" and stampedes were a constant threat because of the electricity in the stormy air. The grass grew too fast and consequently was coarse and "washy" so that the steers, far from gaining weight as they should have on a properly conducted drive, actually lost instead. Another factor hit the market hard. Many of the cattle sold in 1872 were purchased by feeders with the idea of fattening them for sale to the packers. These fat cattle, plump on good corn, were just coming on the market as the first of the 1873 trail herds arrived from Texas. Packers and private butchers alike naturally preferred the corn-fed beef to grass-fed rough steers, and the market for range cattle was very sluggish.

Slowly the summer dragged and by September most of the trail herds were still being held in the country about Newton, Wichita and elsewhere, unsold and apparently unwanted. And then, on September 19, the famous "Black Friday" of 1873 broke, prostrating Wall Street and sending a panic through the nation. Cattle prices plummeted with other values. Perhaps, indeed, the livestock interests of the West suffered more than other branches of commerce in America. Of the period McCoy wrote:

For many weeks, to be upon a livestock market was, to one in sympathy with the operators, like witnessing a daily calamity. So depressed was the business, and so severe were the losses sustained, that whole days would be passed without one being able to hear a lively or jovial remark or see a smile upon the universally sad and gloomy countenances of the dealers.

By scores, in that trying year, the cattlemen went bankrupt, and numerous banks and other financial institutions connected with the livestock industry likewise failed. But, then as now, cattlemen possessed amazing vitality. The ranch industry was quickly on its way back up.

Discovery, in 1874, of gold in the Black Hills of South Dakota had an important bearing on this. With his long golden curls and his circus uniform, Custer led an expedition into the hills and his announcement of his findings started a stampede to the forbidden area—which was guaranteed under a treaty with the Sioux—and launched the Sioux war of 1876. The most notable victim of the Sioux fury in that war, as we have seen, was this same Custer, the cause of the fighting.

But by the day of the Little Bighorn, hordes of gold seekers, undeterred by the government's half-hearted attempts to keep them out, had penetrated the district. Among the mines discovered was the famous Homestake, foundation of the Hearst family fortune and today still the largest producing mine in the nation. Every train going West was crowded. Space on stage coaches going into the hills was at a prohibitive premium. Many rode horseback or even walked. Among those drawn from the cattle country were Wild Bill Hickok, greatest of all gun-fighting experts, and the man who shot him in a Deadwood saloon while his back was turned—Jack McCall.

At once, owing to this sudden concentration of population, there was a great demand for beef. Young men spurred their horses on the Laramie and the Chugwater, rounding up herds for the mining towns—Deadwood, Rapid City, Custer, and their roaring sisters. Business boomed in Cheyenne, which still turned its back on cattle and regarded itself, like Denver, as a mining supply depot with its destiny bound up with the gold fields. Even in Montana, far to the north and on the other side of the Sioux country where scattered hostile bands roamed long after the main bodies of Indian warriors were defeated, stockmen made ready to dare the dangerous drive to the mining centers.

Before the Indians were swept away and concentrated on their reservations by the Sitting Bull war, the cattle range had been pretty rigidly held south of the North Platte. Within the peninsula between the two branches of the Platte developed the first cow country of any extent in the Northern plains, but even as far south as that the menace of Indian war parties was acute. Every spring, as soon as the chinook winds had started the frost out of the ground, Sioux and Cheyennes swooped down on Southern Wyoming. Three forts futilely tried to protect the ranchers—Forts Laramie, Fetterman, and D. A. Russell. By no means could these posts begin adequately to ward off the frequent raids, so the cattlemen fell back on their own resources. In those days most ranches had walls pierced with loopholes—just as the walls of Texas ranch houses had been pierced during the Comanche troubles. Business in Cheyenne was brisk in rifles, revolvers, and ammunition.

These conditions quickly changed after the final defeat of the Sioux. No sooner were the Indians out of the country than the cattle people began to reach for the rich grasslands of the Powder valley, the Big Horn basin, and farther north into Montana's high plains. At once, in a season almost, and even before the buffalo all were gone, the flood gates seemed to burst. A vast, trampling, bawling tide of cattle, with its sinewy outriders, and its stolidly trundling chuckwagons, flowed resistlessly into the Northwest.

It was in the winter of 1877-78 that this movement began from Southern Wyoming as a sort of advance guard for the greater migration which should come later from Texas. The first herds which felt their way into the Powder and Big Horn basins were preceded by scouting cowmen, prospecting for the best ranges. That spring, when the territorial governor of Wyoming returned from an inspection journey through the northern part of the territory, he met several parties of tanned, heavily-armed men, riding keen-eyed in search of a likely place to throw a herd of cattle. By 1879 enough ranchers were in the Powder River country north of the Platte to organize Johnson County—and to prepare the stage for the Johnson County war.

Meantime, foreseeing the rivalry for stock cattle which the newly opened ranges were certain to create, some of the more aggressive and far-sighted Wyoming cattlemen rode far down the trail to meet the Texas trail drivers, seeking to buy at distant points cattle for new ranches that were springing up everywhere on the Powder, Tongue, Big Horn, Upper Cheyenne, Graybull, Shoshone, and other rivers hitherto unknown to civilization. In spite of this activity, however, one hundred thousand cattle that year passed right across Wyoming to the plains of Montana, where already the cowmen of the little mountain ranches were beginning to push out. By 1883 six hundred thousand cattle were in Montana, far more than that in Wyoming. The Northwest was an integral part of the cattle country.

A great new trail was created from Texas, farther west than any trail heretofore, except the Pecos trail of Loving and Goodnight. Because of the Indians many Texans shunned the Red River crossing which had been used by herds going up the Chisholm trail. It was not that the Co-

manches and Kiowas were on the war path. Emphatically
they were at peace. But they had learned that even peace
had its compensations. At long last the idea had entered
their skulls that the reserve on which they lived was theirs
and that white men entering it without permission were
guilty of trespass. Moreover the soldiers at Fort Sill stood
ready to defend the Great White Father's newly tamed
red children in the possession of their property. This contained some truly delightful possibilities. When, for example, trail herds started across their country, bands of
Comanches and Kiowa warriors met them and exacted
heavy toll of "wo-haws," which the cattlemen paid, knowing the government would back up the Indians.

It became the tendency, therefore, to swing west from
about Decatur, Texas, striking the Red River at a place
which came to be called Doan's Crossing, from which a line
due north carried west of the Wichita Mountains, passed
through old Camp Supply and Dodge City, missing the
worst Indian country, and then broke northwest toward
Ogallala, Nebraska, the first outpost of the northern cattle
country.

Doan's Crossing is worth a book in itself. It was a collection of a dozen buildings on the banks of the Red River,
where the trail herds found an easy way across. Judge C. F.
Doan and his nephew Corwin Doan, former Indian traders
at Fort Sill in the Comanche country, were the original
settlers. Judge Doan—he wore the title by courtesy only—
left the Indian trade to carry supplies to the buffalo hunters, and when that evanescent profession faded away in
the extermination of the bison herds, he and his nephew
opened a shanty store at the crossing which was to bear
their name.

Very soon Doan's was known throughout Texas as the
best jumping-off place for the great route to the Northwest. And the Doans grew prosperous selling supplies,
whiskey, ammunition, guns, saddles, bridles, blankets, and
the hundred and one other things men needed in starting

out on the long and adventurous northern trek. The Western trail was the longest of all the trails. At its southern end it tapped the brush country of the Nueces and Rio Grande, with branches leading from San Antonio and the Eagle Pass region. Another branch from the upper Trinity and Brazos valleys joined the Western trail at Doan's Crossing. Entering the Indian Territory, the route forged northward and was joined by still another major tap line at Camp Supply, bringing into it cattle from the older Chisholm trail. Thence it progressed across Kansas through Dodge City, to Ogallala, where a branch left it, taking cattle to the Laramie plains and the Chugwater country. The northward branch, however, continued to the Black Hills, where it divided, one route turning off into the Powder River valley and the Big Horn basin, while the other continued north to the Dakotas, eastern Montana, and even to the Canadian plains of Calgary. More has been written of the Chisholm trail than any of the other routes, but the Western trail and Doan's Crossing in their day eclipsed that famous cattle highway.

One of the first Texas herds to travel the full length of the Western trail was piloted by James H. Cook in 1876, while the Sitting Bull war still was very energetically being fought. The herd of twenty-five hundred longhorns was made up on the Nueces in southern Texas, and it finally reached its destination in North Dakota on the upper Missouri, where it was sold to the government to supply the Indian agencies of that section. Cook's trail herders encountered all the vicissitudes of high water, storms, hunger, thirst, and stampedes, in their journey which spanned the nation from north to south, but they were very lucky for all that. They could not have known, when they began the drive, that they would be crossing the Indian country while the greatest of Sioux wars was raging. Yet in the months they were pushing their steers across the plains east of the Black Hills, Custer was losing his life, Crook was accepting defeat, and MacKenzie and Miles were feeling for the hostile camps all over the Northern plains.

Some guardian angel must have watched over the little band of Texans, and over their great herd of wide-horned animals. Calmly they made their leisurely way north and never met a Sioux.

How many cattle went up the Western trail nobody knows. A firm headed by Captain John F. Lytle sent some 450,000 head over it in a period of several years. In a single season Colonel Ike T. Pryor drove thirty thousand cattle north, and he kept at trail driving for a dozen years. George W. Saunders, later president of the Texas Trail Drivers' Association, Shanghai Pierce, Seth Mabry, Doc Burnett, Dan Waggoner, John and Ab Blocker, Kokernut and a host of others, whose names are classic in the cattle country, went up the dusty Western trail. Saunders is the only one who has attempted an estimate of total numbers. He said that of the twelve million horses and cattle that went to the northern markets during the trail-driving period, six million crossed the Red River at Doan's Crossing, for the Western trail; five million at Red River Station, for the Chisholm trail; and one million at other points. The sale of those vast herds brought $250,000,000 in money to Texas—a tidy sum even in these days of vast national debts, and a sum which saved the state from bankruptcy and developed the present mighty empire it is.

In the last great series of drives to Wyoming and Montana, however, trail driving as an institution in the cow country was reaching its end. The day of the permanent cattle ranch was at hand. Farmers were taking up more and more land, stimulated by the railroads which had built westward with the subsidy of government acreage and had promoted settlements to stimulate their own traffic. As the tide of granger control moved westward, the trail herds found their route more and more difficult. Taking alarm from the signs in 1884, but taking alarm too late, the cattle-

men from the entire range country met in a giant convention in St. Louis. More than two thousand delegates were
there—representing every corner of the cow country. In
addition to the most notable ranchers, breeders, feeders and
owners in the industry, there were numerous packers,
bankers and other interests represented, with a good many
senators and congressmen also.

Two great questions brought the cattle growers there—
two interests that in a measure conflicted. The ranchers
from the Northwest—Montana, Wyoming and the Dakotas—wanted Congress to authorize the long-time leasing
of public domain range lands for grazing. It was a proposal long advocated as a safeguard for the stockman from
the encroachment of the grangers, and a guarantee of per-
manency. The ranchers from Texas and the Southwest
wanted Congress to establish a national livestock trail, extending from the Red River to the Canadian line, a strip
averaging three miles, and not exceeding six, in width, with
wider grazing areas at certain places, quarantine grounds
where necessary, and crossings for native cattle at stated
points where the trail was to be only two hundred feet
wide. Lands required for the trail—which would have been
six hundred and ninety miles long, with an area of more
than two thousand square miles—were to be withdrawn
from settlement or sale for at least ten years. The trail was
to be fenced, and some enthusiasts even advocated bridges
over the rivers to eliminate the necessity of swimming
herds across.

But here the individualism of the cattle country became
its weakness. The cowmen could not come together. Because the ranchers in the Northwest felt the country was
already in danger of overstocking and did not wish to encourage further trail driving from Texas, the stockmen
from that section refused to support the cattle trail. At this
refusal the Texans would not back the grazing lease proposal. As a result Congress failed to enact either proposal

into a law, although Representative James F. Miller of Texas introduced a bill providing for the national cattle trail.

The time for such action, as a matter of fact, was already past and it is doubtful if it would have carried in Congress, even with the united support of the range country. Farm interests had grown too powerful and the railroads likewise threw their influence against it. One of the objections was that the scheme was too elaborate. It would have required millions of dollars to carry out. It died a-borning.

After that trail driving decreased steadily. By the middle '80's the chief movement of herds north were from the big ranches, such as the X I T, Matador, and L S, which bred cattle in Texas and grazed them in Wyoming, Montana and even up in Calgary. Only a quarter of a million cattle went up the trail in 1886 and the winter of that year brought disastrous blizzards that put an end to the cattle boom. By this time the railroads were tapping the Texas country and the expenses of trail driving had grown so great that the rails might well be used to save time and put the cattle in the Northern pastures early enough to acclimatize them for winter. Nesters and farmers combined to heap charge after charge on the trail drivers—for damages to fences, irrigation ditches, crops, and also from disease spread among their cattle. Local law officers enforced these imposts. Gradually the Western trail swung farther and farther toward the mountains, especially after Kansas adopted a state law prohibiting trail herding across its territory. Eventually the route became constricted to a narrow corridor through eastern Colorado, running from the Texas Panhandle though Lamar, Colorado, to Lusk, Wyoming, and ending at Miles City, Montana.

It is recorded by Edward Burnett that the last herd which went up the Western trail was an X I T bunch, driven through in 1896, with John McCanles as trail boss.

The drive was one continual battle, but McCanles finally delivered his herd.

That ended trail driving. No longer was there need for it. The cost of bringing a herd north by railroad and trail were equalized, and men were weary of the eternal feud with the nesters.

XXII

THE CHANGING RANGE

Even before the trail-driving era ended, a new way of life appeared in the cow country—an entire set of customs and habits and ideas, not in the process of development merely, but already settled, established and traditional. The range had come into its own as an entity in American civilization. Long migratory, the trampling herd spread out over an empire of grassland, became as static as ever it could become in a country open and almost innocent of fences.

In advance of the cattle migration the foresighted had throughout the Northwest pre-empted the best water rights. John Iliff's gigantic seizure of riparian ground on the Platte was duplicated in lesser scope by hundreds of other men and outfits. Wherever a supply of year-around water was found contiguous to plentiful grazing, ranch houses sprang up. After the custom of the range, in universal force since it was established in the days of Cortés, cows could graze whither they listed on land that was free to all. But water was not free. Wherefore arose the practice, not new to be sure, of line riding—with cowboys on horseback patrolling imaginary lines which divided their outfits from those next, "throwing back" cattle not belonging to them, and keeping the cattle of their own brands from wandering. On their rounds the line-riders frequently met and exchanged news and cigarettes. But on occasion riding the line became an arduous, even a perilous, occupation, particularly when drouths caused the disappearance of water on some unfortunate range and the cattle sought to find drinking holes on land not controlled by their proper owners. Such periods often were tense. By the very nature of his calling the cowboy was callous, but he was not by instinct or inclination cruel. To prevent his neighbor's cattle, dying of thirst, from drinking at this water hole was not due to malice or meanness, but to the necessity of self-preservation, because range and water usually were not inexhaustible and all were needed for his own stock.

Sometimes the harsh exigencies led to shootings, even to small range wars, and it is to be recorded that no matter how good friends were the cowboys in the rival camps, their loyalty to their own brands and outfits overruled all other considerations. Many a lean rider, forced by circumstances, caught his sights against the breast of a boon companion held in affection, but just then fighting on the other side. The cowboy pulled his trigger with regret, but his regret did not prevent his sights from being steady or the trigger squeeze scientific.

Texas, mother state of the cattle industry, had furnished the seed herds for the Northwest as well as the Southwest; she also furnished the technique of the range. Wyoming and the Dakotas learned how to manage their stock directly from Texas, absorbing also the cattle language, and, indeed, many of the Texans themselves. Today the talk on the Powder River and the Big Horn closely resembles, in its slow, half-humorous drawl, the talk on the Brazos and the Pecos. Traditions and customs are more Southern than Northern.

Montana, on the other hand, learned the cattle business from Wyoming, which had learned it originally from Texas. This was owing to the types of country involved. Wyoming, with its wide plains, took naturally to the cattle methods worked out through long generations by the Texans. But Montana ranchers first began running cattle in little secluded mountain valleys of the Rockies, hidden away from prowling Indians and fenced off by crags from wandering buffalo herds. Chiefly their cattle were from Utah and from Idaho, which in turn drew considerably from the Pacific coast. There is a vast difference between mountain ranching, where every valley is a natural enclosure, and ranching on the plains where there is nothing to prevent an animal from wandering in any direction it desires. When Montana cattlemen came out on the plains, therefore, they had to unlearn some of their own cattle

methods and acquire brand-new ones. It was from Wyoming that Montana obtained its introduction to the chuck-wagon, the cow camp, the open country roundup and other institutions of the wide grasslands.

Geography somewhat changed the language of the cow-men in the Northwest. Where the influence in Texas was Spanish, the influence in Montana and Wyoming was French. Thus the mesa of the Southwest was a butte in the Northwest. Similarly arroyo became coulee, sierra changed to mountain, llano to plain, and rio to river. The mustang found itself referred to as a cayuse or fuzztail. The Spanish *jaquima* was corrupted into hackamore, and lariat, which itself was a corruption of *la reata*, was plain rope in the Northern plains. Such pure Spanish words as *morral* and *frijoles* were lost entirely in the Anglo-Saxonisms, nose-bag and beans. *Vaquero* became twisted into buckaroo, and the Mexican word *cocinero*, for cook, wound up as coosie by the time it reached the Platte.

Essentially, however, cow ranching in the Northwest, for all its geographical differences, remained much as it was in Texas. The Northern cowboy continued his work on lines laid down centuries before by the half-Indian *vaqueros* of Spanish America, who taught it to the white cowmen on the Nueces and Rio Grande, who in turn passed it on to the riders of the upper Missouri and the Yellowstone. The chuckwagon, which had proved its in-dispensable qualities on the trail, found an important new status in the roundup. It was modified somewhat because no longer did it need to be so large and strong. Rarely was it necessary to carry water between roundup camps, so the water barrel often could be dispensed with. But the rear compartment for storing food and utensils and the end-gate working table were as of yore, and so were the Dutch oven, the giant coffee pot, the bedrolls, and the cook—right down to the dilapidated flour sack he wore as an apron and the quid of tobacco which generally bulged his bristly cheek while he mixed the biscuits. The menu was much the same, too—biscuits and gravy, pieces of fried meat, and strong black coffee, with no butter, cream or sugar. It still remained almost a violation of the range code to eat your own beef, if beef of another brand—and hence, according to superstition, sweeter and more nutritious—could be obtained. Nor did the outfit which without a by-your-leave furnished beef to its neighbor look upon that circumstance with an eye too invidious. The instincts of hospitality and generosity were strong in the West, and besides, sooner or later the opportunity would come to repay the compliment. In the end, it was held, all things evened themselves.

Local variations in outfit made themselves apparent. There was the matter of the saddle. This most important of items in the cowboy's outfit varied according to personal taste. Saddles, to one who has used them, are soon seen to possess individualities. After you "break in" a saddle it seems to fit you better than it does anyone else. Its seat and skirts develop hollows and bumps to complement the contours of the rider in much the way that a pair of shoes adjusts itself to the feet. Since he spent most of his waking life in the saddle, it was of the utmost importance to the cowboy that his saddle be comfortable. When once he had broken a saddle in, therefore, he became very loath to part with it, and the instances are many in the West when a reckless youth rode to town, sat into a gambling game, and lost horse, six-shooter, cartridge belt, money—everything except his saddle, which he carried on his back when he trudged back home to the ranch. "He's sold his saddle," is a cattle country phrase. It is capable of an infinity of interpretations. If a man has betrayed a trust or done something else to earn the contempt of his fellows, "he's sold his saddle." If he has lost his business and become destitute, "he's sold his saddle," but with a note of pity rather than scorn. If his mind is deranged and his mental powers gone, "he's sold his saddle," with still a different inflection. And so on. The words imply the ultimate in the abandonment by fate of a cow country person.

A stock saddle is a heavy, handy article. Who invented it nobody knows, but the Spaniards brought it to America, and were themselves introduced to it by the Moors. Medieval Spanish saddles in museums look much as do range saddles of Western America. Before this type of saddle became Moorish it was Arabian, and perhaps even Persian. Almost everyone is familiar with the appearance of the stock saddle; with its deep seat; its horn, formerly of wood but now of steel, making a peg strong enough to "dolly" a rope around and throw the heaviest range bull; its broad hanging skirts; its latigo straps; its wide wooden stirrups;

and its tapaderos. But there were variations even in so standardized a piece of equipment. The California tree was "center fire"—an affair with a single cinch. The Texas saddle had two cinches. In the Northwest, and also in the El Paso-Albuquerque area, where Texas and California met, so to speak, a third style—a hybrid—appeared, known as the three-quarter rig. Eventually all these styles mingled as to locality and each man indulged his own taste.

As to other details of equipment, the buckskin fringes on chaps disappeared, sometimes altogether, sometimes— particularly in recent years—being replaced by the familiar broad "wings," often ornamented with silver conchos. From over the mountains also came another distinctive style—the angora skin chaps, with curly hair outside; a good type which sheds rain and cold as well as protecting the wearer in thorny bush. Boots continued to be close-fitting and high of heel, for the value of this apparently frivolous style was proved—the heel preventing the foot from slipping through a stirrup when a horse was bucking and tapaderos were not worn. The fate of him who is thrown and dragged by a foot caught in a stirrup is well known. Those high, sharp heels, moreover, made excellent snubbing brakes when roping from the ground.

In universal use, also, remained the wide hat, its broad brim a protection from either sun or rain. Many styles in "sombreros" were affected, and the method of creasing, pointing, crushing, or rolling would, at one period, have unerringly revealed the section of the cow country from which the wearer hailed. But in the old days the crown of the hat was low or pinched in, and the high "ten gallon" style of the modern Hollywood cowboy was all but unknown.

Life and custom were stabilized throughout the range country, again with minor variations according to locality. For example, a bunk house in the Southwest was likely to be of adobe, the thick walls being excellent to ward off the desert sun, whereas there was little rain to wash or crumble the material. In the Indian Territory the bunk house might be of logs, loosely put together with little regard to chinking, because the weather was rarely of extreme severity. But in Wyoming and Montana, the bunk house would be well and strongly built, the logs chinked carefully, because the temperatures there sometimes dropped far below zero, and of warmth in the edifice there was great need.

Men still wore brush jackets and heavy tapaderos when they rode in the brush country of Texas. The practice of fastening the hat under the chin with a string belonged also to the brush country, where a low branch might sweep a hat away at any second. In the wider, more open plains, the open stirrup was common and the chin strap less common.

"Parson" Barnard, an old cowman who was by no means a parson, but who did write a valuable book of memoirs called *A Rider of the Cherokee Strip*, gives a summary of the cost and essentials of an average cowboy's outfit in the early '80's:

1 good saddle	$50.00
1 bridle	7.50
1 pair spurs	5.00
1 quirt	2.50
1 lariat rope	2.00
1 slicker	3.00
1 J. B. Stetson hat	10.00
1 hatband	2.00
2 pairs California blankets	20.00
1 sugin	2.00
1 wagon sheet	5.00
1 pair boots	12.00
1 cartridge belt	2.50
1 box .45 cartridges	1.00
1 cow pony	50.00
Total	$189.50

Observe in the above summary that the two essentials of the cowboy's life cost the most—the horse and saddle. In Barnard's list they are rated equally at $50 each, but actually the saddle frequently far outcost the animal that bore it. Kirk Askew of Kansas City, whose saddlery firm formerly furnished more saddles to the range country than probably any other, remembers making saddles for $500 and even $1,000—the latter show affairs or prizes. The $100 to $150 saddle was common. To the trade, fancy frills put on saddles were known as "millinery." Askew has told me that his firm used to make their saddles all alike as to tree, horn and cantle, then wait for orders before putting on the rig and millinery. They would learn the locality for which the saddle was intended, and rig it—center fire for California and Nevada, double cinch for East and Central Texas, and three-quarter rig for the Northwest and the Southwest beyond the Pecos—unless there was a specific order for some other kind of rigging. The amount of leather tooling, conchos and other gew-gaws, was limited only by the price the purchaser wished to pay.

Horses, on the other hand, frequently were relatively cheap. Many a cowboy cinched a $150 saddle on a $20 horse, nor saw anything incongruous about it.

Everywhere, from Canada to Mexico, the roundup existed and still exists in the cow country. And everywhere in its essentials it was the same, with a method and technique worked out patiently by generations of cowmen from the day when Gregorio de Villalobos brought to the New World his first little bunch of calves. The roundup partook of the characteristics of the farmer's harvest and the business man's inventory. It was the time when the beef herd was gathered for the market, when the increase was enumerated, the losses of the previous year noted, and the strays of other ranges returned to their proper places, while lost cattle were received back from distant areas.

In sections where brands were mixed, several outfits often worked together. Four successive steps comprised the full process of the roundup: gathering the cattle, cutting out the desired classes, branding, driving to market the beef herd.

Early in the morning, long before dawn, "coosie" was at work by his winking campfire, preparing the morning meal. A good cook was a treasure, and a bad one bred discontent. Men the world over enjoy good food and cowboys have always possessed Gargantuan appetites. A shiftless, lazy cook would condemn his outfit to a monotony of sour-dough or soda biscuits, sowbelly, white gravy and black coffee. An average cook varied his menus somewhat by adding canned corn and tomatoes, frying beef and roasting ribs, or serving stewed apples. But a really accomplished cook would give his outfit something to brag about with such a gustatory array as: son-of-a-gun (Mexican kidney stew), county attorney (a stew made from veal), Charlie Taylor (a substitute for butter made from a mixture of sorghum molasses and bacon grease), flap-jacks, soda sinkers—and of course the usual array of fried beef rolled in flour, canned tomatoes or corn, beans, stewed dried apples or apricots, and the inevitable black coffee. Such a coosie was a jewel without price, the pride and joy of his entire outfit, and was guarded jealously against the blandishments of other outfits who might seek surreptitiously to lure him away with offers of sordid lucre.

Absence of milk or cream in a cattle country was always remarkable to a newcomer. Cowboys were horsemen and not farm hands in those days, and still are, to a degree, in most of the cattle country, although the bucking seat of a mowing machine has become as much recognized as a cowboy's throne as the saddle of a broncho, and he wears rubber boots in the irrigation ditches almost as frequently as his cowman's boots in the corral. But dairying was absolutely beyond the pale. A rancher who sought to induce his men to milk cows had shortly to seek new men. Joe

Cross, one of the grand old cattlemen of the Panhandle, who still lives on his ranch near Troup, Texas, says that when he was a boy he had to obtain some milk for his family. It took a good man with a rope to snub a long-horn cow to a snubbing post and both hind feet had to be tied. The formula worked out by Cross for range cows was: four gills make a pint, two pints make a quart, and four longhorn cows make a gallon.

A few years ago John E. Roberts, manager of the X I ranch near Meade, Kansas, had unexpected guests. Roberts was a college graduate who had specialized in dairying and knew the niceties, and he was humiliated that with four thousand cows on his range he did not have cream for the coffee of his visitors. In the late evening while the guests were in another part of the ranch he called his cowpunch-ers together, explained the crucial situation, gave each a tin cup, provided one with a bucket, and sent the dozen or more men riding out to bring back a gallon of milk. By roping and throwing cows and squeezing out a cupful or less from each one, the cowboys finally, after a prodi-gious amount of swearing and sweating, filled the gallon bucket with milk. In some triumph they started back for the ranch house, rubbing their bruises. But when the man assigned to carry the bucket mounted his horse, that ani-mal, never having carried a man with a bucket before, very promptly objected. With a squeal he "swallowed his head" and began to buck. It did not take long for the rider to bring him out of it, but by that time it was too late. About the second crow-hop every drop of the hard-won milk was spattered over the prairie. Roberts took one look at the faces of his men as they rode back and discreetly refrained from further suggestions along that line.

But back to the cook fire which we left at dawn: As the first pale streaks of mother-of-pearl in the east gave presage of the coming sun, the long howl, "Come an' git it!", from the cook aroused the sleeping men. Swaddled lumps stirred on the ground, the cowboys rolled out of their blankets, trussed up their bedrolls, tossed them in the chuckwagon and quickly crowded about the fire, eating breakfast before going to the rope corral to catch up a horse for the morning.

Usually some of the steeds were fractious that early in the day and the riders, mounting them, had a lively few minutes "taking the pitch out of them." This was very much a part of the day's routine, and when the bronchos were quieted down, with directions from the roundup boss given, the men rode out across the country, spreading out to sweep a broad area. Having reached a specified limit, they turned and began to bring all cattle toward the common center near the camp. This was far easier to do on open plains than in brush country, but it required hard riding anywhere—and an utter lack of caution or regard for safety of neck and limb. Prairie dog villages, coulees or arroyos on the arid plains, have a way of appearing without warning squarely under the nose of a running horse, with the ever-present prospect of disaster if the animal makes one mistake in its footing while going at its full speed.

A full day might be required before all the cattle in the section to be worked were gathered in the central herd. That night guards would patrol the borders of the herd of sleeping animals, singing, in the manner learned during trail-driving days, their dolorous cowboy melodies. After breakfast the next morning the "cutting" or "chopping" began. This is the separation from the herd of cows with unbranded calves in preparation for branding and tallying. Now the prize horses of the range had their chance to show to beautiful advantage. Nothing prettier was there to watch in the West than a fine cutting horse in action. He knew more about handling cattle than did his rider in some cases. There are instances when, to settle a bet, a prize chopping horse was ridden into a herd, shown the cow to cut out, then left to his own devices, his rider sitting motionless in the saddle and permitting the animal beneath him to do the maneuvering without guidance, until the particular cow selected was safely out of the jam and into the cut herd. The sharp turns, sudden bursts of speed and sharp, jolting stops required in such work were hard on the riders, but they were doubly hard on the horses which had to be changed two or three times a day during a roundup. One thing was worth noticing. Almost all cowboys used severe bits, but so light were their hands that it was rare that a horse's mouth was punished. Most cutting horses turned at the touch of the reins on the neck, sometimes at the mere leaning of the rider's body to one side or the other.

After the cows with unbranded calves were separated from the other cattle, the beef herd, if one was desired, was cut out and the rest of the animals permitted to drift back to the range. Roping and branding of the calves now began, always a spirited and sometimes an amusing sight. A good calf roper went about his work as methodically as if he were not displaying a highly specialized and difficult art. Sitting in his saddle he swung his lariat, perhaps only slightly, and dropped it over the neck or about the front or hind legs of a calf almost at will, in spite of the other cattle crowding about.

In the brush country the lariat was short—rarely more than thirty feet—but elsewhere it was longer, and in the open plains sometimes fifty or sixty feet. There were various types of throws and fastens, and the critical eyes of the cow country judged instantly how expertly they were executed. In the Southwest each had its name, which J. Frank Dobie has gone to the trouble to list. The deadliest throw was called "forefooting." The noose caught the animal by the forefeet and spilled it. In the Southwest this was known as the *mangana*. This style is particularly effective for calf roping, and no sooner is the rope cast than the horse turns for the branding fire, the rider shifting over his weight to one stirrup so his body does not come into the line of the rope, and the calf is dragged, sliding along on its belly, to the branding crew. Dobie lists variations of the *mangana*, including the *mangana de pie*, in which the opened loop was laid on the ground, with one of the thrower's boot toes under the honda, and pitched at its target by a kick as the horse or steer to be roped came by; the *mangana de cabra*, thrown with a twist which made a figure 8 out of the loop, each of the smaller loops catching, one the neck and the other the forefeet of the animal which fell when the forelegs were drawn up to the neck; and the *pial*, aimed at the hind feet of the animal, the two loops of the "8" each catching a separate foot which were thus twisted together with no possibility of kicking loose.

Some of these throws, to one who has never seen them, sound impossible, but not only are there cowboys and *vaqueros* in the Southwest who can do them, but even little Mexican boys, *pastores* of herds of goats or sheep, are skilled in their use.

When the calf was roped and brought to the fire, action was sharp. On the little animal pounced two men called "flankers." One doubled up a front foot and put his knee on the calf's head. The other braced a foot against one hind leg and stretched the other hind leg to its full length. In this position the calf was helpless and almost motionless. The brander set his hot iron on it, the cutter castrated and ear-marked it, and the checker tallied it on the roundup book.

Sometimes a calf was roped about the neck and had to be "hoolihanned" or "bulldogged" down. The calf-wrestler caught the rope near the neck of the animal, reached across its back and took a good handful of the loose skin under its body. As the calf jumped, the man shoved a knee under it, which, aided by a strong upward lift on the opposite side, by the handhold of loose skin, tripped the animal and caused it to fall heavily. Usually its breath was knocked out by the treatment and it lay still long enough to be stretched out and branded.

There was a sufficiency of hard work about branding, whether on the open prairie or in a corral. Modern ranches are usually equipped now with chutes and stanchions, which greatly simplify the work and take much of the labor out of it. But under any circumstances the sun is usually blazing hot, the dust mounts in choking clouds, there are many kicks and tumbles in handling the animals, and the smell of blood, burned hair and hide, and perspiration is over everything.

XXIII

BRANDS AND BRAND BLOTTING

The West had an apothegm: Only three things merited a man's continuous respect—a cowboy who could "stay awake all summer and catch up on his sleep in the winter," a good ranch cook, and a cattle brand. Of course there were also a good woman, a good two-gun fighter, and a good horse—but these were held to be gifts of God and not to be classified with mundane things.

The practice of branding is almost as old as the race and has by no means always been confined to the marking of livestock. Egyptian inscriptions dating as far back as 2000 B.C. indicate that the Pharaohs branded their cattle; also their slaves. The early Greeks burned a Delta on the cheeks of their bondsmen, and the Romans seared an F for *fur* (thief) on the cheeks of convicted robbers. As late as 1830 galley slaves of France were branded. Cortés, conquering Mexico, as we have seen, burned a G for *guerra* (war) on the cheeks of Aztec captives he sold into slavery. And the story of *The Scarlet Letter*, by Hawthorne, tells of the New England custom by which the brand was worn sewn on the clothing, but it probably had its inception in a letter burned on the skin.

Branding of cattle and other livestock has, however, flourished most extensively in America. Cortés, who first branded human beings on this continent, introduced also the brand on cattle when, after the Conquest, he settled down and became a *ranchero*. His cattle brand was three Christian crosses and it was the first in North America. Since that day branding has become almost universal in the cattle country of the West and has developed into an authentic heraldry, the brands serving as the escutcheons

243

of a wild aristocracy and its retainers, as did the coats of
arms of the medieval knights—except that for the Lion
Couchant, Fleur-de-Lis, Griffon, and other chivalric em-
blems of feudalism, the range country substituted down-
to-earth symbols like the Hog Eye, Turkey Track, Walking A, and Buzzard Rail.

As in medieval heraldry, the branding system has its
conventions. A single letter, for example, can have numerous mutations. The letter W, given a spreading tag at the
top of each arm becomes a Flying W; with rounded angles,
a Running W; inside a square, a Box W; with an angle
above it, a Rafter W; with a half-circle under it, a Rocking
W; and so on in an almost infinite variety of changes
which can instantly be recognized and "read" by anyone
familiar with brands.

Few brands, however, are single letters. The purpose of
a brand not only is to mark the animal, but to mark him
so that a rustler will be unable to change it. Various complicated and colorful designs are the result. One of the most
famous brands was the X I T, standing for "Ten in Texas,"
because the huge ranch extended into ten Texas counties.
Another was the Hashknife, a brand which had much to
do with the Tonto Basin war. John Blocker's Block R was
known everywhere and so was John Chisum's Long Rail
and Jingle Bob, which has previously been described.

Some interesting brand names are: Pigpen, Mustache A,
Bible, Crutch, Cinch Buckle, Spanish Bit, Anvil, Swinging
L, Hat E, Anchor, Crazy E, Stirrup, Apple Bar, Dinner
Bell, Andiron, Scissors, and Forked Lightning S.

One brand was called the "Grab All," or "The Hell."
The latter name came from the fact that the two words
could be worked out of its design which might be described
as a Lazy H and Flying E connected. It was run for years
by Looney & Prude, in Jeff Davis County, Texas. Colonel
Frank D. Hatch's brand, the Forty-five, was another spectacular brand. The old colonel was a fire-eater, and he
always said that he designed the brand "so that everyone

will know that the owner of Forty-five cattle is apt to shoot first and find out what is the matter after the smoke clears away."

Some cowboys were proud of their ability to read brands, and occasionally controversies arose over the proper designations of certain ones. Such a situation occurred over the brand known as the Satchel or the Cotton Hoe, depending on the "reader." Cowmen used to become pretty warmed up debating the correct reading of that brand. Another peculiar mark was Matthew Cartwright's, which can be described as a Flying Half-Circle Diamond and a Half. The Flying Half-Circle was sometimes called the Straddle Bug, so the brand name could be simplified down to Straddle Bug Diamond and a Half. But a name as lengthy as that didn't suit the cowboys. They dubbed it the Fleur de Mustard, and that is the way it stands.

In west Texas there was a brand which defied even the most practiced brand readers. It belonged to the late Henry M. Halff and consisted of a half-circle, open side down, above another, open to the left. Forty years ago it was a noted brand in Midland, Upton and Glasscock Counties. When it was first used, a Mexican *vaquero* who had some reputation as a brand reader, was asked to name it.

"*Quien sabe?* (Who knows?)" he replied after studying it. Ever since the cowboys have called it the Kinsavvy brand.

The De Vaca family, which claims descent from Cabeza de Vaca, first of the explorers of the Southwest, has since earliest times used a symbol representing a *cabeza de vaca* (cow's head). It is still being run in Arizona.

Some of the bigger outfits did not deem it necessary to have complicated brands, depending on their size and power to keep off rustlers. The famous King ranch, today the world's largest, uses a simple Running W; and the Matador Company has the distinctive, but equally simple, Matador V.

As important as the brand on the animal's flank or side

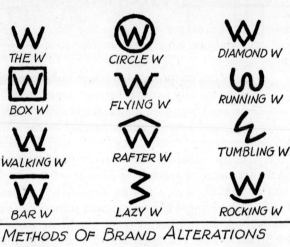

THE W
CIRCLE W
DIAMOND W

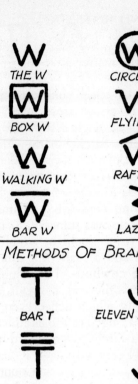

BOX W
FLYING W
RUNNING W

WALKING W
RAFTER W
TUMBLING W

BAR W
LAZY W
ROCKING W

METHODS OF BRAND ALTERATIONS

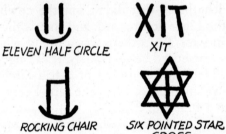

BAR T
ELEVEN HALF CIRCLE
XIT
XIT

CURRY COMB
ROCKING CHAIR
SIX POINTED STAR
CROSS

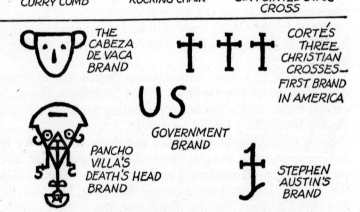

THE CABEZA DE VACA BRAND

CORTÉS THREE CHRISTIAN CROSSES— FIRST BRAND IN AMERICA

US
GOVERNMENT BRAND

PANCHO VILLA'S DEATH'S HEAD BRAND

STEPHEN AUSTIN'S BRAND

BRAND VARIATIONS ON ONE LETTER

HASH KNIFE	BLOCK R	SWINGING L	PIG PEN	MUSTACHE A	BIBLE
CRUTCH	CINCH BUCKLE	SPANISH BIT	ANVIL	ANCHOR	CRAZY E
STIRRUP	SATCHEL OR COTTON HOE	APPLE BAR	QUIEN SABE	DINNER BELL	FLYING E
RUNNING OPEN A	ANDIRON	FORKED LIGHTNING CROSS S	SCISSORS	DIAMOND AND A HALF	HAT E
BUZZARD ON A RAIL	HOG EYE	WALKING A	TURKEY TRACK	XIT	LONG RAIL
GRAB ALL OR THE HELL	WINE GLASS H	FORTY-FIVE	FLEUR DE MUSTARD	STRADDLE BUG	B TRIANGLE
CRUTCH BAR	MOON	OX YOKE	FOUR SIXES	MATADOR V	FRYING PAN

SOME NOTABLE CATTLE BRANDS

is the ear-mark. This is for the purpose of preliminary identification. When you ride up to a herd of cattle, they may be crowded together so that they obscure the flanks of one another; but every head is turned inquiringly in your direction with the ears forward. Those ears are the most quickly apparent features of a cow. Suppose your cattle are branded on the side with a Rafter A, and ear-marked with an under bit on the left and a slope on the right. When you approach the herd you cannot determine from the brands how many of your cattle are in it, unless you inspect separately each animal—a manifest impossibility. But a glance at those ears will permit you quickly to pick out your cows, to be verified by the brands, and thus will reveal those with calves at foot for branding.

Some of the ear-marks used are: Crop—the ear cut squarely off. Under bit—a nick in the lower edge. Over bit —a nick on the upper edge. Over slope—an angling crop on the upper edge. Under slope—the same on the lower edge. Swallow fork—a triangular piece cut out of the tip. Jingle bob—the ear split to the head, letting the pieces flap. Split—the ear divided less fully than the jingle bob.

Another marker is the jug-handled dewlap, cut in the fleshy part of the throat.

None of these minor markings were accepted as final proof of ownership. They were for quick identification, after which the ownership was proved by the brand. Rustlers, working on the open range, often took advantage of this circumstance. They would rope calves and ear-mark them with the rightful owner's mark, but they would not brand them, trusting that the ropers would overlook them in the roundup, noting by their ears that they had been handled and assuming they were properly branded. Next season, after the calves were old enough to leave their mothers, they were known as "sleepers," and the rustlers roped them at leisure, branded them with the rustlers' brand, and gave them new ear-marks of such nature that the former insignia was obliterated.

Not all rustlers took the trouble or time needed for the "sleeper" plan. Many wanted quicker results. No matter how carefully designed a brand was, it could usually be altered by a clever cowboy, so that its original markings would blend into the new design. For example, a letter T with a line above it is called a Bar T, but if another line is added above, it becomes a Curry Comb. One of the celebrated brand blottings made a victim of an outfit which ran a figure 11 over a half-circle, called the 11 Half-Circle. The rustlers extended the two lines of the 11 downward to meet the half-circle, connected the tops with a bar, drew one of the lines farther up than the other, and produced a brand they called the Rocking Chair.

Another famous alteration was on the X I T. The brand blotters worked out a six-pointed star with a cross in the center, taking advantage of the fact that the up-and-down bar of the T usually was a bit out of line. Surprisingly, the difficult X I T was burned into another blot—the 4 Box P. One of the yarns of the Panhandle is that a cowboy who had been involved in the six-point star blot on the X I T brand was arrested and tried at Lubbock, Texas. The jury took a look at the X I T and promptly acquitted the accused on the ground it was impossible to change the brand as alleged. Then, when he was freed, so the story goes, the rustler demonstrated how the brand could be made— on a bet. He hunted around in one of the X I T pastures until he found an animal on which the up-and-down bar of the T, through carelessness, was a little out of line. On this he made his alteration.

One of the hardest of all brands to alter was Burk Burnett's Four Sixes. The 6 6 6 6, according to tradition, came into being one night when Burnett sat into a big poker game and was phenomenally fortunate. Another rancher was losing heavily and eventually Burnett won everything the other possessed except his ranch. Another hand was proposed by the losing cowman—his ranch against the pot, double or nothing. Burnett accepted the challenge and drew four sixes, to win. He took for his brand the lucky poker hand.

In states where a brand law exists, a brand must be properly registered to denote title. There is a story of a Texas man, evidently a greenhorn in range customs, who obtained and branded a herd of cattle but neglected his registration. A dishonest neighbor discovered the oversight and one night ran off the entire herd; then registered the brand in his own name. Although the original owner protested that the cattle were stolen from him, he was powerless before the courts, so strong were brand law and custom.

In the West the big ranchers were always carrying on a war against cattle thieves. Sometimes the latter used a "cold brand" or "hair brand" in misbranding. The rightful owner's brand was either put on through a wet blanket, or else so lightly pressed that while it burned the hair down to the skin it did not scorch the animal's hide. Except on close inspection it looked like a legitimate brand, but in a few months, when the calf had left its identifying mother and its hair had shed, the brand disappeared completely and the rustler could rebrand at his leisure with any brand that appealed to him.

John Chisum is said to have tried to trap rustlers on his range with a decoy brand, a small, concealed brand between the legs which he placed on several large maverick steers he scattered about at his water holes. Unfortunately the first cowboy who put a rope on such a steer found the decoy brand, saw through the scheme, and promptly burned on the animal a large question mark. Another cowboy saw this strange insignia, wondered what it was, roped and threw the steer, and, discovering also the decoy mark, added another brand of his own. After that, it is said, the unfortunate steer was roped every few days and each time was given a new brand until his hide had the appearance of one of those collegiate slickers of a few years ago.

But there was one cattleman who solved the brand blotting question. He was Pancho Villa, the Mexican bandit chief. Villa trusted nobody. And he evolved a brand bearing the appropriate name of the Death's Head. It was so complicated that although the cattle which bore it numbered many thousands, nobody was ever known to have successfully altered it.

Out of the constant depredations of the rustlers, as well as the common legislative and other problems besetting the cattlemen, grew the cattle associations. One of the first of these was known simply as the Stock Growers' Association and was organized in Colorado in 1867. It changed its name, in 1876, to the Colorado Cattle Growers' Association, to keep it distinct in the public mind from the sheepmen of the state.

The Colorado association became the parent of many others. Numerous smaller groups sprang up in local areas of Kansas, and the Cherokee Strip association long waged a successful struggle against throwing open that section to farm settlement. The Montana Stock Growers' Association was organized in 1884 with the famous Northwestern cattleman, Conrad Kohrs, as the moving force in it. It took in the Territory of Dakota, and thus acquired another member who became even more famous than Kohrs. He was a young rancher named Theodore Roosevelt.

The notable Wyoming Stock Growers' Association was formed in 1873 and achieved a position of such political power that it elected many members to the territorial legislature and controlled the state government for years. Possibly the greatest of all cattlemen's associations, however, is that of Texas and the Southwest. It was organized at Graham, Texas, in 1877 as the Stock Growers' Association of Northwest Texas. From the first it was successful and a few years later united with the South Texas Association under the name of the Cattle Raisers Association of Texas. Eventually it consolidated with the Panhandle and Southwestern Stock Raisers Association, whereby it took the name it now bears, the Texas and Southwestern Cattle

Raisers Association, including in its membership not only ranchers from Texas, but from other states as well. The American National Live Stock Association is the national organization of cattlemen today.

One of the present and past services of the big cattle associations has been the system of inspectors maintained at the big stockyards over the country. No more competent individual exists in the cattle industry than the quiet, poker-faced man under the cowboy hat, who sits his pony in the Chicago, Kansas City, Denver, Omaha, Fort Worth or any other major stockyards and "tallies off" the brands on the rumps of the steers as they run down the chutes.

That man in the big hat has a memory that makes a card-index system appear to be a victim of amnesia. The cattlemen out on the range owe much to his quick eye and memory for brands. Sweeping over a trainload of steers, his glance instantly detects the brands of animals that have intruded from another range, or cattle that have not been branded at all. Not only does he know the brands registered with the association he represents, but most of the other brands in the West, and at his challenge any stockyards will hold up the sale of one animal or a train-load until the ownership is properly established.

At one time there was a considerable furore on the part of humane societies and kind-hearted old ladies in the East over the supposed cruelty of branding.

As a matter of fact the process of branding, ear-marking and castrating calves apparently is not very painful to the animals. They will bleat and struggle when they feel the hot iron, but the knife affects them little. As soon as it is released, the calf is taken in charge by its mother, and within a few minutes it will usually be observed feeding, which would not be the case if it were suffering pain. About the most agonizing process to cattle is dehorning in the old style, with a stanchion and saw. The head is held immovable while the saw rasps its way through the tender nerve centers at the base of the horn. A steer will almost tie himself in a knot during this process and I have seen full-grown animals faint from pain and shock while being dehorned. Fortunately dehorning now is usually done with clippers, and in the case of calves up to three or four months old, with a gouger, which is an instantaneous and far less painful process.

At the time when the East was protesting loudly against branding, and the West was giving no heed, some society in New England or elsewhere suggested that instead of branding the cattle little metal disks should be stamped with the owner's name and address, and affixed with wire —like a sort of earring—in the calf's ear. The well-intentioned proponents of this plan were probably utterly astonished and mystified at the immoderate laughter which greeted their scheme when it was heard out in the cow country.

One other celebrated attempt was made to evade the necessity of branding. In the spring of 1887 Finlay Dun, a Scot with no experience in American cattle ranching, took over the management of the Swan Land & Cattle Company of Wyoming, of which he was a director and large stockholder. Dun proposed to revolutionize stock handling methods. Instead of branding, he ordered paint by the barrel. On twenty-nine thousand amazed and frightened cattle he had his paint daubed, but Wyoming sun and rain made short work of his humanitarian experiment. Even Dun had in the end to admit his defeat while his cowboys—behind his back—chanted:

> Oh, Daddy Dun's a dandy,
> But his paint won't stick.

In his report to his company Dun stated with preciseness that the paint was not "sufficiently adhesive and permanent."

There is no question that branding is the best method

yet devised for marking cattle when handled in large
numbers in an open country. Barbed wire fences caused a
decline in it in the heavily populated areas, but the West
has clung to it. Last year the legislature of Kansas, after
decades without a state branding law, re-enacted one as
the best safeguard against the new type of cattle rustler,
who operates today by truck and along paved roads.

XXIV

TO WHILE AWAY THE TIME

Close to two of the immeasurables of the universe—space and time—lived the cattle country. Man and his affairs were dwarfed into insignificance by the immensity of the plains and mountains. Time moved with stately inevitability, the days marshaled into months, the months into seasons, and the seasons into years. Men reckoned back by roundups, except that certain years were marked by special disaster, as the year of the big blizzard, the grasshopper year, the year of the cholera epidemic, and other elemental dates which remained in the folk consciousness of the West— and still do, for that matter.

It is not, however, to be supposed that ranch life was an unrelieved round of hard riding and hard work. Slack periods were frequent, and when the weather was fine and the men loafed about the bunk houses, it was instinctive to look for amusement. In any primitive society the instinct for gambling is strong. To the cowboy it was the great relaxation. Horse racing was a passion in the West, but most cowmen needed no formality for betting. A wager on which of two birds, sitting side by side on a fence, would fly first as they were approached by jogging horsemen; on who could spit the most accurately at a knot hole or a grasshopper; or on any trifle of the kind would be offered and instantly accepted.

To cards the cowboy took naturally. Mexican monte, seven-up, faro, black jack, pitch, cribbage—he exploited them all to their uttermost. Poker, however, was probably his favorite.

If poker is the great American indoor sport it is so because of the refinements made in it by Westerners—cowboys and miners—who evolved a large share of the game's fine points. Intensely American as the game now is, poker probably originated in Persia where it was played by four persons using a deck of twenty cards. Later it found its way to France and Germany, where it was known, respectively, as Poque and Pochen. By English corruption the form poker was evolved.

Tradition has it that a group of young men from New Orleans, sons of planters who were sent to Europe to complete their educations, brought the game back with them to Louisiana. Its psychic possibilities appealed to the American mind. Up the Mississippi it traveled with the river gamblers as its chief tutors—at great profit to themselves —until the public began to learn its mysteries, when the contest became progressively less one-sided. From the great river, poker went east and west—out on the plains and into the populous areas along the Atlantic. And from the first it was embraced by the gentlemen of the South as their peculiar pastime.

Not all at once was poker evolved as it now is known. It was, for instance, about 1830 before the full deck of fifty-two cards was used. Even then, for another thirty years, it remained a simple game called "bluff" as often as poker. Five cards were dealt face down to each player and he bet on that deal. There was no draw. The primitive ratings of the card combinations were, starting at the lowest and going to the highest in regular order: high card, one pair, two pairs, three of a kind, and four of a kind.

When the War between the States broke out in 1861 and hundreds of thousands of men in the army camps sought some form of recreation to take their minds off the daily strain of the fighting, the convention of the draw after the first deal was introduced in poker. About the same time the straight was added as a card combination and shortly afterward the flush. These two combinations at first ranked above two pair but below three of a kind. As evidence of the often friendly communication between the men of the opposing ranks in the War, the rules of poker circulated freely in the camps of both the Blue and the Gray, and were more or less standardized by common consent by the end of the conflict.

The release of the soldiers after Appomattox sent thousands to the frontier to find new homes and new occupations, and they took with them "the old army game." Out on the plains, in the early era of trail drivers, buffalo hunters and Indian fighters, the game of stud poker, considered the present-day standard, was formulated. Science then really entered the game. Heretofore the combination rankings had been guesswork, more or less, but some mathematical genius worked out the tables of probabilities and Mr. Hoyle's invaluable compendium embalmed the rules in deathless print.

The cowman was a natural and enthusiastic poker player. To a great extent he is responsible for the national enthusiasm for it, and out of the game has come a substantial part of the rich American language. Think of these phrases: "bluffer," "four-flusher," "lay the cards on the table," "cash in," "showdown," "passing the buck," "penny ante proposition," "ace in the hole," "ace high," "stand patter," "calling a bluff," "cold deck," "sit in," "turn down," "rotten deal," "square deal," and even "new deal"—they are all poker terms, and most of them from the range country.

In the ranch bunk houses a game was pretty constantly going, and on the trail cowboys played for small stakes. But when the cowpuncher went to town with his pay burning his pockets, he generally first sought a bar and freely libated, then took a seat at the nearest poker table and plunged with his earnings until he went broke or "struck it big." "Double or nothing" was his credo.

Much has been made by the motion pictures of singing and crooning cowboys. Singing was a part of the cowboy's life. He crooned or sang to the sleeping herd on night watch, and he developed a rich repertoire of salty songs, most of them unprintable. Sometimes cowboys even sang around campfires, as the films have depicted them. But the cowboy who rode about yawping loudly to the moon without a good legitimate reason would have been regarded with suspicion and distaste. The cowboy might secretly plume himself on having a good voice, but he did not take a stance on an imposing rock pinnacle and burst into song for the admiration of his fellows, nor did he carry a guitar on horseback while he was herding his cows. That particular type of exhibitionism was not for the cattle country. Singing on the range was usually reserved for a time when it would not draw the most bitter, acrimonious criticism the language of the West could concentrate on it—and that was when the necessity of work demanded it. I understand that nowadays crooning cowboys are in demand on the dude ranches, but cows and debutantes have different standards of music and I doubt if the real songsters of the old cow country would have pleased the romantic soul of a modern girl tourist. As for the old-time cattle—one of the modern swing-time songs would probably have set a herd of them off on a forty-mile stampede.

Practical jokes sometimes were played in cow camps,

but they had to be indulged in with some circumspection, because a joke taken the wrong way might very easily result in gun-play with unpleasant, if not fatal, results. Yet every cow town had a group of loafers who spent their time thinking up ways to put the treats on one another, or on newcomers; and swapped jokes, spun yarns, and aimlessly discussed the weather or cattle prices. Out of the ennui of these town indolents rose an odd craze that swept the West in the late '70's and early '80's—the Lazy Men's Clubs. Hardly a Western town but had one at some period in its history. Reference to them can frequently be found in the files of old newspapers. The rules generally prohibited members from doing any kind of labor, physical or mental, and those who violated the code "stood the treats." At Jacksboro, Texas, a man was convicted by the club of "assisting, by looking on, while another man laid the floor of a house." Another conviction was on the charge that a member "borrowed a newspaper and read it himself."

Women always were few in numbers in the cattle country and their place was correspondingly high. Every trail town, every mining camp, every railhead, of course, had its population of prostitutes—some bedizened, shrill-voiced and brazen, others occasionally of great beauty and attractiveness—with whom the cowboys were all too familiar. But when there was a "respectable ball" in which only "good women" took part, the masculine element attended from long distances in the country around. Frequently there were only half a dozen women to forty men. Under such circumstances no woman could be a wallflower, no matter how unattractive she was. If she were young and pretty the rush for her as a dancing partner was almost alarming. One mirthful young lady from the East attended a dance in Texas, and afterward wrote a spirited and amusing account of her impressions, which was printed in the

Lubbock, Texas, *Avalanche*. Parts of it are worth re-reading:

It was with many misgivings in spite of my partner's assurance that he would pull me through, that I took my place in the dance.

> "*Hark ye partners,*
> *Rights the same.*"

So far, I bowed as did the rest.

> "*Balance you all.*"

With the plunge of a maddened steer, my partner came toward me. I smothered a scream as I was seized and swung around like a bag of meal. Before I could get my breath I was pushed out to answer for:

> "*First lady to the right;*
> *Swing the man that stole the sheep,*
> *Now the one that hauled it home,*
> *Now the one that ate the meat,*
> *Now the one that gnawed the bones.*"

Not being acquainted with the private histories of the men in the set, I was at a little disadvantage, but I was seized, swung, and passed on to the next, until I finally arrived breathless at the starting point.

> "*First gent, swing yer opposite pardner,*
> *Then yer turtle dove.*
> *Again yer opposite pardner,*
> *And now yer true love.*"

I blushed in spite of myself at so publicly passing as my partner's "turtle dove" and "own true love," while his sweetheart over in the corner transfixed me with a jealous glare and saw no humor whatever in the situation.
Again the command:

> "*First couple to the right,*
> *Cage the bird, three hands around.*"

I found myself in the center of a circle formed by my partner and the second couple and then exchanged places with my partner at the call:

> *"Birdie hop out and crane hop in,*
> *Three hands around and go it agin.*
> *All men left, back to the pardner,*
> *And grand right and left.*
> *Come to yer pardner once and a half*
> *Yaller hammer right and jaybird left,*
> *Meet yer pardner and all chaw hay,*
> *You know where and I don't care,*
> *Seat yer pardner in the old arm chair."*

By this time, feeling quite bruised and battered, I was ready for almost any kind of a chair.

The leather-lunged "caller" whose crude versifications are above described was typical of the Western dances, and so were scraping fiddles, stamping high-heeled boots on the rough floor, and a general fine fury which imbued the whole affair. The West was young, full of life and spirits, and it craved above all things else, action.

At the rear of the dancing room, or in the kitchen, often stood a whiskey barrel, and a beef was frequently barbecued to refresh the guests, some of whom rode as much as a hundred miles to be present at a merry-making. Otherwise, however, the demands on hospitality were not great. Everyone brought his own bedroll and his sleeping chamber was the illimitable out-of-doors.

They were holding such a dance as this at Fine Ewing's Mule Creek ranch in the Indian Territory one night in the '80's. The cowboys far outnumbered the girls, and those young men who weren't dancing had recourse to the whiskey barrel so frequently that some of them were more than a bit unsteady when dawn came.

On the previous night a train with a stock car, containing fifteen buffalo consigned from Charles Goodnight's ranch on the Palo Duro to some zoological park in the

East, stopped at Kiowa, Kansas, where the bison were unloaded to be watered and fed. In some manner the beasts broke out from their pen and started southwest for the home range, picking up cattle which joined the stampede as it went along.

Just as the bibulous cowboys were emerging from Ewing's ranch house, the wild herd went thundering by, led by the buffalo, and with three or four hundred steers behind frantically trying to keep up. It was just light enough to make out the outlines of the bison and surmise the whole herd was stampeding buffalo. An apparition out of the past! There had not been a wild bison herd in the Territory for more than a decade. Dode Peters, a U cowboy, who had been drinking heavily, fainted. He never could be convinced after he was revived that he had not seen a vision inspired by the fan-tods due to too much liquor.

Crude as were those frontier dances, they gave the cowboys an opportunity to meet the few unmarried girls of the cattle country. There were bashful advances, giggling replies, and the comely girls married quickly, youth being strong and lusty and normal in its instincts. Usually a big ranch consisted of the owner's home, the foreman's home, a bunk house for the men, a cook shack, stables, corrals, and so on. Quite often the rancher was a man of family. So, not infrequently, was the foreman, and as the range country grew older, some of the more settled cowpunchers began to adventure into matrimony. It followed that there were children in the cattle country, and these played, as do children the world over, at the pursuits of their elders. Little boys grew up with ropes in their hands. From the moment they could swing an infantile loop they practiced throwing it, and life was a burden to every dog, chicken, calf, or colt that came within reach. Miniature corrals were built and imaginary cattle branded. Rustlers were battled and Indians scalped. Down in the real corral, boys backed hard-bucking little calves and began to acquire the racial skill in riding which remained with them through life.

For the elders these things at which the children played were matters of sober, every-day business. Yet it is natural that when a man is skilled at anything, he should pride himself in it and practice it for his amusement. So the lariat was in frequent play on things other than cattle.

No complete record ever has been compiled of the different creatures that have been roped. I have seen enumerated buffalo, mustangs, grizzlies, black bear, lobo wolves, mountain lion, coyotes, deer, antelope—an amazing feat considering the animal's speed—wild turkeys, bob cats, *javalinas*, elk and even African lion. But the two top roping feats on animals that have come to my notice are as follows:

The late Uncle Dave Cummings of Buffalo, Wyoming, told me that he had personally witnessed the roping of a prairie dog. A cowpuncher did it on a bet, placing a bait in the form of an apple some distance from the hole so that the dog would have to come out to nibble it. The man lay in wait behind some sagebrush. He swung a small loop with a thin rawhide rope and got the dog right behind the front legs before it reached the hole. It sounds impossible, but that's the story.

Even more surprising was the exploit of José Romero, a Tucson cowhand, who actually roped and captured a full-grown golden eagle from horseback, May 31, 1939. Romero rode up on the bird while it was eating on a carcass. Before it could rise he whipped his noose about it, and so became possibly the first man ever to put a rope on a wild eagle.

It will be gathered from the foregoing that the rope was a source of sport as well as of labor. When the cowboys gathered for the roundup they sometimes competed with one another in trials of skill, vying in long-fasten and fancy throws. Riding bad horses was another form of amusement and every outfit had its champion bronk twister. Out of these simple, informal sports arose the rodeo, a combination of a circus and an athletic contest, which still has its enthusiasts all over the country, including Madison Square Garden, New York.

One more roping record should be mentioned. It was made by Ellison Carroll a quarter of a century ago, when he won the title of world champion roper. He tied a single steer in sixteen seconds. Then, to prove it was no accident, he roped and tied, one after another, a carload of twenty-two head of wild Texas steers—for an average time of twenty-two and one-half seconds per steer. Mr. Carroll is still living at Big Lake, Texas, and still, so far as I know, he has to see a successful challenger of that record among the younger rodeo stars.

XXV

THE HORSE A MAN RIDES

There is no doubt that the first horses in America were Spanish, landed by Cortés when he began the Conquest—unless you wish to include those prehistoric ancestors of the horse whose skeletons are dug up in the fossil beds of the West to ornament our paleontological museums. Spanish, also, were the first horses to enter the United States, those ridden by the soldiers of Coronado in his abortive expedition. But in spite of his derivation, the Western cow pony is a true American from nose to hocks.

The original Spanish horses were of Arab or Barb blood, the highest types of equine beauty, speed, intelligence and endurance. So long had their strain lived in the tents of the desert nomads that they developed instincts concerning man which persisted after generations of living virtually untamed in the wilds. On the Spanish blood was superimposed the consequences of a stern battle for existence in the American wilderness, so that after four centuries the Texas mustang and his more domestic though scarcely tamer relative, the broncho of the cattle ranches—lineal descendants both of the Spanish Arabian—were peerlessly fitted for cow-hunting in the brush, trail-driving across the wilderness, and roundup work on the high plains. In the centuries of his arduous wandering, it is true, Nature subtracted from the mustang something of the beauty and grace of his progenitors, but she did not impair one whit his intelligence, and as for stamina and reserve strength, they were immeasurably increased.

The mustang, roaming the plains, breeding and increasing, became the greatest revolutionizing influence in the life of the plains Indians. Almost overnight, when he learned to ride a horse, the Indian changed from a foot skulker to a swashbuckling paladin. The Comanches probably obtained horses first, but other tribes quickly imitated those bowlegged corsairs of the plains. Even in primitive times, however, travelers noticed a difference in breeds of horses among the Indian tribes. The horses of the Sioux, for example, were larger and stockier than those of the Comanches, owing perhaps to the more abundant and nutritious forage of the Northern pastures and the greater severity of Northern winters.

After the white man drove the Indians on reservations and took over the plains, the Western broncho remained for generations a perfect working animal for the cattle range, until within recent years his blood was chilled by the draft and driving horses of the white settlers. Chief criticism of the broncho was that he was small, and the demand grew for larger horses. As a result numerous strains were fused with the mustang breed, but at least one high authority, Dan D. Casement, one of the West's best known stockmen, is convinced that the gain in size, weight and substance was more than offset by losses in endurance, handiness and cow sense.

In Oregon heavy Clydesdale sires were turned on the range with mustang mares. The progeny were short-set, compact, with big bones and hairy legs. Though willing and strong, these horses were awkward, and the cowboys, who loathed them, called them "Oregon lummoxes." Over in Montana and the Dakotas, Percheron sires were similarly tried. The get of these stallions, while larger and somewhat more graceful than that of the Clydesdales, were still big-footed and awkward, and earned the name "Percheron puddin' foots."

In the Southwest, however, and particularly in Texas, a different type of horse was evolved. The Southerner loved a fast horse. By selection and breeding he produced what was known as a "quarter horse," an animal with great speed from a standing start for short distances up to a quarter of a mile. It was the opinion of the late William Anson, of Cristobal, Texas, that the famous Justin Morgan, progenitor of the Morgan strain, was a quarter horse. In Texas, however, the most famous sire was Steel Dust, brought to the Lone Star State, according to tradition, from Kentucky for the specified purpose of a match race

against a locally celebrated champion. The local horse had vanquished so many opponents from the surrounding hinterlands that the simple denizens of the cow country could not believe it was not invincible and wagered their money with such prodigality on the race that the subsequent victory of Steel Dust impoverished the entire district. But a delegation of earnest gentlemen with hardware on their persons waited on the importers of Steel Dust and remonstrated against any attempt to ship the horse back out of Texas.

"Ye got our money, an' we ain't begrudgin' it," these

gentry said with firmness. "But now that hoss is in this yere country, we ain't aimin' to let it leave."

Steel Dust remained in Texas and his get was numerous and of high quality. Through this great sire and others of his type, the Texas pony became a faster, better rounded, smoother horse, with small, alert ears, well-poised front, short, deep barrel, and powerful quarters. Better still, this type of cross-breeding did not dull the cow sense and stamina of the offspring. Horses from strains like this were in great favor with the cowboys. They were the "Sunday horses," the special "chopping" horses of the roundup.

The endurance of the Western broncho is legendary. Definite proof of the little plains horse's staying powers was contained in a test made by the United States government in 1897. Dr. William A. Bruett, special commissioner of the bureau of animal industry, that year arranged a twenty-four hundred mile ride from Sheridan, Wyoming, to Galena, Illinois. Two young cowboys, Bill and Bert Gabriel, made the ride. From a ranch near Sheridan they caught up two bronchos, neither of which ever had worn a saddle or bridle, broke them, and on June 5, 1897, started for Illinois. Neither horse had eaten a mouthful of grain in its life and neither knew what it was to be shod. The route covered all types of country—turf, sand, rock, clay, and mud. When the bronchos reached the end of their two thousand, four hundred mile run at Galena on September 6, having been on the road ninety-one days during which they had subsisted entirely on grass by the wayside, they were in condition as good in every way as when they left Wyoming.

The purpose of the test was to demonstrate the value of native Western horses for army and export use. The Gabriels' mounts were true bronchos, weighing 750 to 900 pounds, in distinction from the so-called range horses, bred up with Clydesdale, Percheron and Hambletonian stock, and weighing from 950 to 1,150 pounds. Their performance actually created a foreign demand for Western horses. In 1893 only five hundred head of the breed had been sold in Chicago. In 1897 the number jumped up to fifteen thousand exported to foreign buyers, chiefly for army use.

There is a difference between the wholly wild mustang and the semi-domesticated broncho. The wild mustang has practically disappeared in Texas, except perhaps in the Big Bend county, but he still may be found in Arizona, Nevada, Utah and Montana—in fact, wherever there are still large areas of wild and unfenced country. Old-timers have said that even when mustang herds were thickest on the plains ranchers preferred to buy gentle horses even though the price was high, and that the Comanches would ride or walk for days through a country containing hundreds of herds of mustangs, in order to steal at great risk a few tame horses from the settlers.

This was because the mustang was extremely hard to capture and still more difficult to tame. Before it could even be trusted to graze without picketing or hobbling it usually had to be removed from its home range. Yet in the early days mustang pens were to be found all over Southern Texas. They were used in trapping the mustang— whose name, incidentally, is from the Spanish *mesteño*, the catchers being known as *mesteñeros*. Professional *mesteñeros* built their pens in clumps of trees or brush so the outlines of the trap could not be seen. Invariably the corral was circular, because trapped horses crowd into corners and either break the fence or damage themselves. The fence was strongly built by digging a deep circular ditch, planting heavy posts upright in it with dirt firmly tamped about them and their tops lashed together with rawhide thongs. Wide wings formed an immense V leading into the gate. These were of brush or tree tops, one wing extending half a mile or more, the other about half as far. At the throat of the V yawned the gate into the corral, built so it could be quickly closed by shooting bars into place. In some cases a self-trap was used. Posts and fence of the corral were concealed by brush and green foliage.

At the start of the drive one rider was stationed at the outer end of each wing. The rest of the horsemen located a mustang herd and chased it, circling the animals toward the longer wing of the V trap, where the posted rider turned them down toward the gate. At this the rider from the shorter wing "pointed in" hard, and kept the mustangs crowded against the wing. By this time everyone was following fast, whooping and hazing the horses into the pen. A man dismounted, threw a bar or two into the gate and hung a blanket over them—which effectively kept the mustangs away from that part of the corral.

The trapped horses were roped from outside and dragged through an opening made by removing posts in the fence. Until they were thrown and tied mustangs usually fought crazily. Several methods were employed to break them so they could even be herd driven. Sometimes they were "necked" to gentle horses or even to stolid burros. Occasionally heavy wooden clogs were fastened about the front legs just above the hoofs, which flapped and caused pain to the horse when it moved faster than a walk. Some mustangs were "side lined," the front and rear legs on the same side being fastened together by a short length of rope. Others were hobbled. Almost always in this process the wild horses were badly bruised and injured before they could be handled. After being driven a few days from their accustomed range, however, they generally became gentler and rapidly recuperated from their hurts.

Other methods of capturing wild horses were "walking them down," in which relays of riders kept a herd on the move night and day until it became so exhausted it could easily be captured; and even "creasing," by firing a bullet so that it grazed the top of the horse's neck and stunned it long enough to let it be hobbled. "Creasing," when successful, was usually a lucky accident. More horses have had their necks broken than were ever captured by it. *Mesteñeros* realized about $2.50 a head on their unbroken mustangs, the broken animals bringing prices somewhat better.

Every horse country has its horse stories, and there are many tales in the West about notable mustangs. Probably the most famous was the white pacer which became the basis of Ernest Thompson Seton's splendid story, *The Pacing Mustang*. It was a pure white stallion, a natural pacer, which ranged in the days of early Texas between the San Antonio and Nueces Rivers. The horse could pace away and leave any pursuer, never breaking his gait.

Many tried to capture him, but all failed, until at last the "walk down" was used to capture all the white stallion's mares, and guile was resorted to in taking the pacing mustang himself. A severe drouth scourged the country and the *mesteñeros* fenced in a lake, leaving open a gate which *might* lead to water, but in reality opened into a pen. Almost dead with thirst, the stallion blundered into the trap at night, was captured, and later taken East, where record of him was lost.

It is the contention of some old horsemen that this white stallion was not a true mustang, being the colt of a white mare lost by some Kentucky soldiers under General Zachary Taylor, when he was marching through Texas to invade Mexico. However that may be, the white pacer had the courage and spirit of a mustang. Other stories exist of white, black, sorrel, even pinto mustangs—an average of one for every locality—but this seems to be the only one where the horse was captured.

In one respect the mustang probably has not had his equal in the equine world. He was the wickedest bucker ever put under a saddle, if he wanted to buck—which was frequently. His short body, steel-spring legs, and sloping rump all were designed for bucking and he possessed a malignant cunning and devil's courage to back up the pyrotechnics he displayed.

In the later years of the West, when the rodeo reached its peak of popularity, many outlaw horses became famous. Whistling Annie, Hot Foot, Smithy, Long Tom, Lightfoot, Casey Jones, Bitter Creek, Old Colonial, Wiggles, Crooked River, and McKay were some of the noted buckers, and men like Lee Caldwell and Jack Sundown, two of the greatest riders of the white and red races respectively, made their reputations on them.

Earthquake, who died just a short time ago, on June 9, 1939, at Shawmut, Montana, was one of the most potent buckers. He was a big horse, and it was his perverse determination not to be broken to a plow that landed him in the bucking ring. Earthquake was sired by a Percheron stallion out of a mustang mare. He was bought by Leo Cremer from the disgruntled farmer who could not make a plow horse out of him, and became famous. When he was nineteen years old he still had enough bucking prowess to throw Burel Mulkey of Salmon, Idaho, who later became world champion cowboy. At twenty Earthquake retired and died at twenty-six.

Probably the greatest outlaw of them all was famous Midnight. Midnight was foaled on Jim McNab's Cottonwood ranch near MacLeod, Alberta. He was raised as an ordinary cow pony, but when for the first time they saddled him, he showed such explosive zeal in twisting off his riders that eventually he found his way to the rodeo circuits. He was only three years old then. For fourteen years his hurtling body and lashing feet grounded the nation's best riders. His owners, McCarthy & Elliott, freely pitted him against the top busters of the West, and he made such men eat dirt as Earl Thode, Pete Knight, Doff Aber, Eddie Woods and Paddy Ryon—all of them names to conjure with in the Western riding world. For years Midnight performed in cities throughout the United States and Canada. Finally, in 1933, when he was seventeen years old, he was ridden twice at Forth Worth. Immediately his owners retired him. Almost with tears in his eyes Verne Elliott announced that Midnight was sick at Fort Worth and would never be entered again.

"He bucked off the best cowboys of his day," said Elliott. "I'm not going to let every hick in the country ride him now that he's sick."

So Midnight was taken to a ranch near Johnstown, Colorado, and lived there in pampered ease until his death, at the age of twenty, in November, 1936. He was given a burial, and a tombstone was erected at his grave just as if he had been a Kentucky Derby winner—or a Christian.

The horse a man rides—or doesn't ride—can come mighty close to his heart.

5. . . . . HIGH TIDE ON THE HIGH PLAINS

XXVI

POPULATING THE PANHANDLE

For decades the *Llanos Estacados* of Texas were a barrier to settlement. A strange, abrupt upthrust of flat plain, which breaks sharply away from the eastward levels, with an escarpment sometimes extremely steep and quite high, it is criss-crossed with canyons and gullies, and so dry that early explorers called it "the Sahara of America." After Texas had poured her longhorns north and west, filling up almost all of the interior basin, the Staked Plains of the Panhandle still remained an immense island of unoccupied territory.

This was so partly because it had been so long the habitat and stronghold of the fighting Comanches, and partly because water was almost lacking. Partly, too, because of the general inhospitality of the flat, treeless expanse. The great cattle trails avoided it, except for the Horsehead route which cut across its southern tip. Other than this the nearest trail to the Staked Plains on the east was the Western trail which skirted its escarpments, and on the west the Goodnight-Loving route up the Pecos. Between these was a huge area of grassland with no human claimants after the Indians were driven off it.

So it might have remained for years longer than it did, had it not been for a panic that brought financial ruin to Charles Goodnight. In 1873 Goodnight was thirty-six years old and newly married. A natural athlete, he stood an even six feet tall and weighed more than two hundred pounds—all hard muscle and bone. Unlike most other cowmen he wore a short-cut beard, and in the resolute lines of his face, the shape of his nose and forehead, and the set of his head on his shoulders, he has been likened to a much larger edition of General Grant, to whom he was not inferior in intellectual qualities or daring.

When the panic of '73 struck the cattle country, Goodnight was ranching prosperously in Colorado. Thither he had taken his young wife, the daughter of an aristocratic Kentucky family, and shortly after her arrival in the rambunctious town of Pueblo, she awoke one morning to find two very defunct bandits hanging peacefully from a telegraph post. These happened to be members of the Coe band of outlaws who had been caught and lynched by the Pueblo vigilantes, of whom Charles Goodnight was an active and aggressive member. Molly Goodnight was horrified and shocked. She suspected, moreover, that her husband had something to do with decorating that post, so she called him to account.

"Well," replied Goodnight, "I don't think it hurt the telegraph pole any."

Afterward he remarked with a tinge of wonder that his words seemed to irritate her, "and it took quite a spell of being nice to her and making her acquainted with the good people of Pueblo before she got over it."

The Goodnights were progressive citizens with strong social instincts. At Pueblo they subscribed to a church and toward a fund to found a college. Goodnight invested in city property as well as ranch lands and joined in the organization of the Stock Growers' Bank in the fall of 1873— just in time to be heavily involved in it when the panic struck.

The bank failed and Goodnight was wiped out as had been hundreds of other stockmen. In the end, his ranch, money and property all were gone, except for eighteen hundred longhorns he had managed to salvage. Some men would have been broken by discouragement, but not Goodnight. With some faithful cow hands he left Colorado and pointed his cattle southeast to the last great open area he knew—the Staked Plains. As a young man he had campaigned with the rangers against Indians in the *Llanos*. Now he remembered the immense carpet of buffalo grass and knew that if he could find anywhere in it a supply of water he might recoup his fortunes.

At first his outfit headquartered on the Canadian River. That was in the Comanche country and, when he reached it in the summer of 1875, the soldiers of Miles and MacKenzie had just finished chastening the allied tribes. Most of the Indians were on reservations, but there were still bands of irreconcilables out on the plains, and working cattle on that range was very dangerous. Goodnight saw to it that his men were well armed, continuously watchful, and of the type that would fight if need arose. Leigh Dyer was his foreman, Dave McCormick one of his best men with rope or gun, and John Rumans another. Two members of the outfit deserve more than passing notice. One was James T. Hughes, whose father was the British author of *Tom Brown's School Days*, and the other was J. C. Johnston, a Scot, who like Hughes rode with Goodnight for the experience. Johnston later was a director of the Matador Company. The two Britons had some money and Goodnight sorely needed capital. So much did they enjoy range life that they invested with Goodnight, buying a third interest in his herd.

A strange thing took place that winter. Down the Canadian valley came drifting several sheep outfits. The herders asked to be allowed to hold their sheep in the vicinity of Goodnight's cattle outfit for protection from renegade Indians. A natural antipathy exists between cowmen and sheepmen, and Goodnight shared the feeling, but he was generous. He marked certain limits of range for his cattle—the first of all sheep deadlines—and told the sheepmen they might remain near enough to obtain aid from his men if they would not transgress those borders. It was a strange alliance and the agreement was violated only once. Then some of Goodnight's cowboys retaliated by stampeding four hundred sheep into the Canadian, where they drowned. That was a mistake. Goodnight later had to pay for those sheep. It developed the transgressing herd belonged to the governor of New Mexico.

Meantime Goodnight still searched for a location on the Staked Plains which lay south of the Canadian. Unexpected help came in the winter of 1875-76. Among the vagrant plains nomads who that winter drifted past Goodnight's camp was a party of *mesteñeros*, out to capture wild horses for the market. With one of them, a man named Martinez, Goodnight had a conversation, and learned he was a former Comanchero, his trade having been somewhat disrupted by the activities of the soldiers a year or so before. More important, Martinez mentioned an enormous wild gorge which gashed the *Llanos* in two, and on the floor of which a city could be built. It was the famous Palo Duro canyon he was describing, out of which in 1874 MacKenzie had routed Lone Wolf's Comanches.

Promptly Goodnight employed Martinez to show him the place. The old Mexican set out with the cowman, but after days of wandering seemed confused by the many other canyons they crossed. Goodnight was losing faith in his guide when at last they rode up to the brink of a tremendous chasm, the genuine Palo Duro, with walls a thousand feet high. From only a few hundred yards wide in some places, the canyon expands to miles in width at others. In the bottom runs Palo Duro Creek, in reality the headwaters of the Red River, with cedars and willows on its banks and buffalo grass carpeting the entire valley. The Mexican clapped his hands with joy.

"*Al fin! Al fin!* (At last! At last!)," he cried.

To Goodnight's purpose the Palo Duro was exactly suited. At once he returned for his outfit and his wife, who was still in Pueblo. The cattle were moved first. They entered the yawning chasm by an old Indian trail down which they had to walk single file in places to the bottom hundreds of feet below. To drive a wagon down was impossible, so the chuckwagon was taken apart and lowered over the cliffs by lariats tied together. At the bottom, once reached, Goodnight's men found an ideal range, but before they could begin to graze their cattle on it they had to rout out the buffalo which occupied it. Goodnight later recounted how his men "choused" the bison down the mighty canyon, firing bullets near the feet of the laggards, until at least ten thousand of the animals were thundering before them, and the little black bears, grubbing along the canyon walls, went scuttling in panic up among the ledges of the rim rock. After that, for two or three years, until the buffalo hunters made it no longer necessary, it was a regular duty of Goodnight's riders to drive back the buffalo from the part of the Palo Duro reserved for cattle range.

It would have been hard to find anywhere on the continent a finer natural location for a ranch. The canyon was already an inclosure and no fencing was needed. Even if the cattle grazed high up its sloping walls the perpendicular cap-rock which ran around the top with a few minor breaks, acted as an effective barrier. Near the creek, on a green stretch of rich level grassland, Goodnight built the log houses of what he called the "Old Home" ranch—the first ranch in the Texas Panhandle. Then he went north to Pueblo for his wife.

On that journey he met a man who was to play a large part in his life—John G. Adair, a wealthy Irishman whose wife was the former Cornelia Wadsworth of the Newport and New York social register. Adair was enamored of the West and at his proposal an agreement was made, whereby Goodnight was to purchase, develop, and manage a ranch, stocking it with cattle and horses, with money furnished by Adair. Goodnight was to receive a salary of twenty-five hundred dollars a year, and at the end of five years was to acquire one-third of the land, cattle and horses, out of which he should pay Adair one-third of the money originally invested.

So sprang up a famous Texas partnership—the J A ranch, the brand of which was made from Adair's initials. This ranch still flourishes. Adair remained chiefly in the East, leaving the management to Goodnight. He and his wife appeared in 1877 to inspect the ranch, but thereafter they seldom visited. Occasionally, however, after Adair's death, his widow would descend upon Palo Duro and give the entire Panhandle a subject for conversation, with her guests of foreign or American social magnificence, her trainload of personal baggage, her battalion of maids, butlers and secretaries, and her luxurious hours of rising in the morning. The widow did not know it, but she was foretasting a much later development of the West—the dude ranch.

It was in the Palo Duro that Goodnight pioneered with experiments toward grading up his cattle. Before the general introduction of fences there was little inducement for a herd owner to improve his strain by introducing pure-bred bulls, because there was nothing to prevent his neighbor from profiting as much as did he by their breeding, since all cows look alike to a bull, whether or not they wear his owner's brand. The Palo Duro canyon, however, was a natural enclosure, and Goodnight introduced high-grade Durham (Shorthorn) bulls. The experiment was unsuccessful, the Durhams being unsuited to the country. Later Goodnight tried to breed up his longhorns by selecting only the finest sires of the native breed. Finally, in 1882, he tried out Herefords and was immediately successful. The Hereford, as it was to develop, was the answer to the Western stockman's problem.

At the end of the first five years the Goodnight-Adair partnership had a profit of slightly more than $510,000 and was continued for another five years. Adair died in May, 1885, at the height of the cattle boom, but his widow continued the partnership with Goodnight. The latter bought more land and at the end of his decade of contract management the J A ranch contained 1,225,000 acres and grazed one hundred thousand cattle, many of which were "white faces" as Herefords became known throughout the West. The ranch today is still one of the largest in the cow country with half a million acres and twenty thousand or more cattle on it.

Goodnight's spectacular success was certain to bring others to the Panhandle. The first "neighbor" was Thomas S. Bugbee who established a ranch "next door"—a mere eighty miles away. Henrietta, Indian Territory, two hundred miles east, was the nearest settlement of any importance. Northeast was Mobeetie, a sprawling of mean shacks with a population largely of gamblers, saloon keepers and ladies of easy virtue. The minutes of the district court, presided over by Judge J. A. Carroll, who traveled eight hundred miles from Emporia, Kansas, to Mobeetie, by train and stage coach to hold his first sitting, contain names that reveal the type of that hamlet's citizens: Frog Mouth Annie, Butcher Knife Bill, Feather Stone Jones, Matilda Wave.

About 1877 the wild little town of Tascosa was founded on the Canadian, at a point where the Comanches and Comancheros had formerly had a meeting place and trading point. Mobeetie and Tascosa at once became rivals and enemies, each with a certain civic pride in the superior brand of its own wickedness. There was, in point of fact, little to choose between them in depravity, but a third town which grew up not far from the J A was exactly the opposite. It was founded by the Rev. Lewis Henry Carhart, and was named Clarendon for his wife. This was a colony of Methodists, farmers rather than ranchers, and preserved a strictly moral atmosphere. The cowboys contemptuously called it "Saint's Roost," but it lived and was protected by Goodnight.

Within two or three years after Goodnight's entrance into the Palo Duro canyon, the buffalo were cleaned out of the Panhandle, except for a small herd Goodnight preserved in an enclosure, and the dam was removed which released a flood of cattle from the great breeding ground of interior Texas. Near Tascosa, George W. Littlefield established his L I T. The Bar C's—a bar above two C's—near Adobe Walls, was Henry W. Creswell's brand. At Buffalo Springs was the Three 7's of Berry & Boice. Other notable outfits were the Turkey Track of the Hansford Land & Cattle Company, the Scissors, Diamond Trail, T-Anchor, and Hat. One-Armed Jim Reed ran his cattle on Tonkawa Creek. The Hashknife of J. N. Simpson was in Taylor County. Other famous names that came to the Panhandle in the first two years after Goodnight were: E. Harrell, the Ikards, Dan Waggoner, Burk Burnett, E. P. Davis, Ike Mullins, R. F. Tankersley, and A. B. Robertson.

Greatest of all the Panhandle ranches was the X I T, established by the Capitol Syndicate and owned by John and Charles Farwell, brothers, of Chicago, who obtained British financial backing and made a transaction with the State of Texas which was typical of the largeness of gesture of the day. The Farwells "traded" a three million dollar capitol building at Austin for three million acres of land on the Staked Plains. The capitol building was erected according to specifications and in its day was a wonder among the nation's state houses. In return the Capitol Syndicate received a deed to the three million acres.

No doubt the state officials of Texas felt they had the best of the deal. At the time the land was considered virtually worthless. The census report of 1880 stated that seven million acres of the *Llanos Estacados* of Texas were absolute desert, unfitted for human use. Within that area was the three million acre tract given the Farwells. In enlarging his J A ranch Goodnight had purchased two hundred and sixty-five square miles—one hundred and seventy thousand acres—for twenty cents an acre. Actually, at the prices then prevalent, the Capitol Syndicate, at a dollar an acre, did receive the worst of the bargain.

But the great tract was laid out—almost four thousand seven hundred square miles of it. It took all or parts of ten counties—Dallam, Harley, Oldham, Deaf Smith, Parmer, Castro, Bailey, Lamb, Hockley and Cochran—which lay along the Texas-New Mexico border. John Blocker, the famous old trail driver, is credited by one of the traditions with designing the X I T brand which stands, as we have noted, for "Ten in Texas," referring to the ten counties. Another story is that John Blocker's brother Ab, with the assistance of Barbecue Campbell, made the design. At any rate it is pretty certain that Ab Blocker—who specialized in a loop so wide that such a loop is now known all over the West as a "Blocker loop"—threw and branded the first steer with it. The statement has been made that the Capitol Syndicate was British owned. This is not true. It was owned by the Farwells, Americans. They did borrow money abroad. A group of wealthy Britons, including the Earl of Aberdeen and the Marquis of Tweeddale, furnished the money that built the state house, but the title of the ranch was in this nation.

By this time the Panhandle had on it as many ranches as could be maintained under its conditions. Not until two things came to revolutionize the entire Western cattle industry would it be fully occupied. Those things were the barbed wire fence and the windmill.

Meantime Panhandle stockmen found themselves grappling with a problem which previously had faced cattle owners in other parts of the country. The Texas cattle tick, which caused all the trouble for the drovers attempting to pass through Kansas and Missouri in 1866, did not cause disease among longhorns, owing to the immunity the latter had built up. But high-grade cattle, which were beginning to appear in the Panhandle, were susceptible to Texas fever. Some of the ranchers, including and perhaps led by Goodnight, began to oppose the trailing of herds of rough longhorns through their territory.

First to take summary action was Goodnight himself. On November 8, 1881, the Fort Griffin, Texas, *Echo* published the following interesting correspondence between Goodnight and another widely known cattleman, George T. Reynolds:

Fort Griffin, Oct. 5.

Editor, *Echo:*

Herewith I hand you a letter which is so plain that it requires no explanation. I desire its publication that stock men generally may know how overbearing prosperity can make a man. Respectfully,

GEORGE T. REYNOLDS.

Que Ti Qua Ranch, Aug. 20.

GEORGE T. REYNOLDS, ESQ.

DEAR SIR:

I send Mr. Smith to turn your cattle so they will not pass through our range. He will show you around and guide you until you strike the head of this stream and then you will have a road. The way he will show you is nearer and there are shorter drives to water than any route you can take. Should you come by here you will have a drive of 35 miles to make.

I hope you will take this advice as yourselves and I have always been good friends, but even friendship will not protect you in the drive through here, and should you attempt to pass through, be kind enough to tell your men what they will have to face as I do not wish to hurt men that do not understand what they will be very sure to meet.

I hope you will not treat this as idle talk, for I mean every word of this, and if you have any feeling for me as a friend or acquaintance, you will not put me to any desperate actions. I will not perhaps see yourself, but take this advice from one that is and always has been your friend.

My cattle are now dying of the fever contracted from cattle driven from Fort Worth, therefore do not have any hope that you can convince me that your cattle will not give mine the fever, this we will not speak of. I simply say to you that you will never pass through here in good health.

Yours truly,

C. GOODNIGHT.

It was the forthright first declaration of the "Winchester quarantine," as the prohibition against trailing became known in the Panhandle. Reynolds did not accept the challenge, contenting himself by expressing his discontent through the *Echo*. Actually, in a more temperate moment, he very probably recognized the justice of Goodnight's position. The first sentence of the last paragraph tells the story: "My cattle are now dying of the fever . . ."

Goodnight was not overbearing or high-handed. He simply took the lead in an unpleasant duty and the first man he opposed was an old friend. The other stockmen took courage from his example. At Mobeetie the Panhandle ranchers met, adopted resolutions in which two trails were designated as permissible for use, and organized the Panhandle Stock Association of Texas. The Panhandle had come of age as a full-fledged part of the cattle community.

XXVII

FINIS TO FREE GRASS

J. F. Glidden, a not too prosperous resident of De Kalb, Illinois, was aggravated one morning in 1874 to discover that his cattle had broken down the smooth wire fence around his pasture. Fencing was always a problem. Stone walls, hedge fences, board fences, picket fences, rail fences —all these were employed. But they were practical only on comparatively small tracts. The cost of fencing large tracts by any of these methods was prohibitive. Which accounts in a large degree for the fact that in the Western cattle country, fences—except for corrals—were almost unknown.

Efforts had been made to use wire. During the War between the States telegraph wire had even been utilized in wire entanglements, foreshadowing the later war-time use of barbed wire. But in 1874 there was no barbed wire; and smooth wire, while permitting an owner to build a fence cheaply, was not cattle-proof, as Glidden discovered that morning.

The De Kalb man began to repair his fence. After a time he noticed, hanging from the sagging wires that had been torn down from the posts, some of the staples he had used to fasten the fence. Twisted about the wire they made little sharp burrs. Glidden began to think. If he could devise some way to place staples at regular intervals on a wire, so they would form rows of spikes, perhaps his cows would have less inclination to tear down the fence. . . .

It was a revolutionary inspiration Glidden had that morning, although at first he thought only of its application to his own problem. He began to experiment with different types of barbs and methods of attaching them firmly to the main wire. At last, using an old coffee mill, he was able to make the coil barb, held in place by wires twisted together—the basic idea behind barbed wire today.

By now the De Kalb man envisaged possibilities for his invention, and on November 24, 1874—an historic date for the West—he took out the first patent covering his device. Later, with I. L. Ellwood, he perfected and patented a machine for producing the new wire in quantities. A factory was built in De Kalb and the manufacture of what was at first called "Glidden wire" began. In its first year the company made only five tons of the wire and sold little of that. A strong prejudice against the "Glidden wire" made itself apparent from the first. Glidden saw that he must overcome that prejudice, and to this end he secured the services of a breezy young friend, Henry B. Sanborn, also a resident of De Kalb.

The solid wall of opposition around De Kalb at first discouraged Sanborn, but he, too, did some profitable thinking. If a market for cheap fencing existed, it must be in the West. That was where the great unfenced areas lay. There ought to be an almost unlimited potential market. Sanborn went to Texas in 1875. On the way he stopped at Kansas City and talked with Major Andrew Drumm, whose huge U ranch lay in the Indian Territory. The result was important. Drumm built the first barbed wire fence of any magnitude on his ranch next year, a fifty-mile drift fence north of his holdings. His example was soon followed by others.

Meantime, all through September, 1875, Sanborn tried vainly to interest the big stockmen of Texas in his new product. A. H. (Shanghai) Pierce, of the bellowing voice and the multitudinous anecdotes, summed up the general attitude:

"It jest won't do!" he roared. "The cows will jest run into that Glidden wire and cut theirselves. Know what that means? Screw worm! Every open cut means screw worm trouble. It would ruin thousands of head in this country."

Undiscouraged, Sanborn telegraphed his home factory to ship a carload of wire to Texas and started an intensive campaign to convince ranchers that their cattle would quickly develop the same sort of caution toward the artificial "thorns" of the new fencing as they observed toward the natural thorns of a hedge. Over the country he traveled by buckboard, visiting ranches and leaving, where he could obtain permission, a few reels of wire to be put up experimentally. The plan worked. Cattlemen convinced themselves of the value and cheapness of the "Glidden wire" and its use began to spread rapidly.

Lumbermen and railroads feared the "Glidden wire" would injure the lumber traffic. With the help of hostile stockmen they introduced a bill into the Texas legislature, making illegal the use of barbed wire. But already the new fencing had taken hold in the cattle country and the bill failed.

In the end the West rushed to fence itself in. Small as well as large ranchers built barbed wire barriers about their grass as fast as possible. As the demand grew and barbed wire became hard to obtain and therefore at a premium, it was not unheard of for a Westerner to "lift" wire from someone else's fence and put it about his own pasture. The top wire only was taken, in such cases, it being esteemed less reprehensible for some reason if the "borrower" stopped at that point. A period intervened, as a matter of fact, when "lifting" wire was practiced by some surprising classes of citizens, and when it was considered, apparently, only a very minor sin.

The late Frank Luther, robust and humorous Democratic power in western Kansas, and long mayor of the town of Cimarron, used to delight in recounting an incident in point. Luther, then a young rancher, needed some wire. There was a long fence belonging to a big cattle company —fair game in any case—and he decided to "borrow" the top wire from a stretch of this.

"I was working along, quietly, up a hill," he would recall gleefully. "I'd pull out the staples for two or three posts, roll the wire to that point, then go to staple-pulling again. Presently I began to notice, once in a while, a tug on the wire I was working on. I couldn't understand it, so I crawled up to the top of the hill and looked over. What I saw plumb amazed me. Coming up the opposite slope of the hill, swiping the same top wire I was working on, was the Methodist preacher of Cimarron's only church!"

One of the first big Texas cattle outfits to use barbed wire was the X I T. Within a short time of the securing of its acreage, it began the erection of an immense mileage of fencing. By the end of two years it had eight hundred miles of wire up and later cross-fences increased this milage to fifteen hundred—enough to reach half-way across the continent. More than one hundred and seventy-five thousand dollars was expended on the outer barrier alone, and so vast was the area fenced that many wild creatures, including buffalo, mustangs, antelope and deer, lived in it for years without even knowing they were prisoners.

Bill Metcalf, a road contractor of the Cimarron valley, secured the contract for building the fence around the huge cattle empire. A man of immense energy, Metcalf marshaled his fencing crews as a general arranges his forces. The X I T requirements were exacting: cedar posts of large size, four strands of wire, the posts to be well set and braced. One of Metcalf's gangs chopped posts in the cedar brakes of the Cimarron and Road canyons and the Carrizo Monte—much of it government land. The posts had to be obtained somewhere and Metcalf, like most frontiersmen, had small scruples on that score. Another outfit hauled the posts and freighted the wire. Some of the cedar logs were brought forty miles or more and the wire came by wagon one hundred and twenty-five miles from Trinidad, Colorado, the nearest rail terminal. Ahead went the post-setting crews, followed in regular order by the bracers and wire-stretchers. Each outfit had its own chuckwagon and cook. Metcalf did a good job for the X I T. Some of the posts he "borrowed" from neighboring government lands are still in use today.

That fence created the first great metamorphosis of the cowboy. He changed from a line rider to a fence rider, and a pair of wire pliers became his insignia of office.

Next a group of small cattle owners along the Canadian organized and built a forty-mile drift fence north of the river to keep back the cattle of the big companies on the plains above. These large concerns, owned by foreign or Eastern capital chiefly, had been extremely high-handed with the smaller ranchers, running roughshod over them at roundup time, and permitting their cattle to drift at will on the already overstocked ranges of the Canadian valley. But if the small owners had wished a peculiarly condign revenge, they could not have hit upon a better idea than that fence. It was built of heavy cedar posts, with five strands of barbed wire. No cattle could possibly break through it. The drift fence proved a death trap in the winter of 1885-86—the year of one of the big blizzards.

The fall was mild and the grass was dry that year. Metcalf was summoned by the X I T boss and asked to plow a fire guard entirely around the ranch properties—a gigantic task, in which hundreds of miles of furrows had to be turned to keep future fires from leaping across and destroying the X I T's big pastures.

In that day a prairie fire was a double disaster. The immediate damage might be great or small. Houses, hay stacks, and sometimes stock or even human beings might be burned. But in the destruction of grass lay the great hazard. After an autumn fire the ground was bare and black for months until the following spring brought new green shoots, and during that period no living thing could subsist on it. Cattlemen went to any length to fight prairie fires. Sometimes fire guards were plowed hastily across the path of the flames to halt them. On other occasions long lines of men—every individual in the country around who

could be summoned—beat and pounded with wet sacking
at the advancing line of flames. In the short grass country
a method sometimes used was exciting and frequently ef-
ficacious. Two cowboys rode with a green, flapping rawhide dragging behind them at the ends of their lariats. On
either side of the line of flames they raced their horses, so
that the rawhide passed directly over the fire, beating it
out. Such a drag often reduced a whole stretch of burning
prairie in a few minutes into sporadic smaller acreas of fire
which were more easily defeated. When, however, a stretch
of tall grass was unexpectedly encountered, or the wind
suddenly rose, the rider on the lee side of the flames might
be in a dangerous position.

Metcalf sent to Trinidad for grain and provisions and
with a huge gang of plows began his contract job about
the middle of November, 1885. Steadily the wide strip of
freshly turned sod scarred its way across the gray-brown
prairie. Paralleling the workers went the chuckwagon.

One afternoon when the fire guard had been driven
miles south along the X I T fence, Metcalf's cook halted
his wagon to prepare the evening meal. Five miles to the
east the plows still slowly drove southward. How it happened has never been explained, but in some manner the
cooking fire caught the prairie grass and began to spread.
A hard south wind was blowing. Panic seized the chuck
wrangler. For a few minutes he tried desperately to beat
out the flames, but they spread wider and wider, traveled
faster and faster. Seeing the hopelessness of further effort,
the luckless *cocinero* collected his personal effects, mounted
bareback on a mule, and fled to Trinidad to escape the
wrath of Metcalf and the X I T.

The smoke of the burning prairie was seen by the men
plowing. They quickly unhitched their mules and hastened
to the camp, but it was too late; the fire was past control.
With an ever-widening front it swept before the swift
wind, burning the entire prairie from the Beaver to the
Canadian. Old cowmen tell of how the entire north end of

the Panhandle and part of the strip known as No Man's Land seemed to be on fire. Many cattle were injured or killed. One trail outfit was directly in the path of the blaze. The men fought fire and extricated most of their herd, but a bunch of about three hundred cows and calves was caught in a stretch of high grass. After the flames swept by the scene was sickening and pathetic. Many of the cows and calves were dead. Others staggered about, their eyesight scorched out. All were badly burned. The very thing the X I T sought to avert had been caused by the cook's carelessness.

Ahead lay still further disaster. Early in January, 1886, after many days of mildness, the great blizzard struck. The cattle did the only thing they could do—they began to drift with the blind smother of white. The Panhandle became a great charnel house. Along the drift fence north of the Canadian were piled thousands of decaying carcasses when spring came. The evil redolence of their putrefaction affronted the nostrils at a distance of many miles away. They were what was left of the cattle of the big ranches north of the Canadian valley. To the drift fence they had gone and there stopped, waited dumbly, and finally went down to rise no more. In the vast area denuded by the prairie fire of the previous November, other thousands starved to death. Snow and fire and barbed wire fences combined to reap a terrific toll.

In spite of its ghastly impost, however, the barbed wire fence had come to stay. With it had arrived a new condition in the range: a man now had need of something more than a few head of cows to go into the ranch business. He had to have capital, because barbed wire and posts and labor cost money. More than ever the little rancher was crowded to the wall, while the great ranches in Texas, the Indian Territory, Kansas, Nebraska, and the Northwest, cold-bloodedly fenced in the water and grass.

Still another factor entered the range picture. All over

the West were huge areas of grazing land which could not be used because, while the grass was excellent, there was no available drinking water. Since earliest Spanish times, of course, the cow country had possessed some water wells. These were old-fashioned, dug by hand, and primitive as to methods of drawing the water, unable to furnish enough to supply even a small herd of cattle.

In the '70's some primitive windmills were brought in at great expense to Texas and used to pump water from hand-dug wells. And then someone discovered that almost the entire plains country was underlaid by aqueous beds—sheets of hidden water that could be tapped at comparatively shallow depths. At once began the great well-drilling boom. Men with drilling outfits went through the Panhandle, the Dakotas, the trans-Pecos, Kansas, Colorado, and elsewhere—wherever water could be found—putting down wells with drill and casing. Upon the horizon the towers and revolving vanes of windmills became familiar features of the landscape. The development came just in time, for natural water was beginning to decrease, owing to irrigation, agriculture and other factors. Today, although surface water is less plentiful than formerly, the cattle country is less dependent upon the vagaries of precipitation than ever in its history. An interesting side-light is the creation of a new occupation on the range. Every large ranch employs a man known as a windmill wrangler. He understands the idiosyncrasies of the whirling wheels that harness the shifting currents of air as well as the horse wrangler understands the movements of his remuda. And because his services are essential, he is a fixture in cattle land.

New impetus to fencing was given by the windmills. In a period of a few years the open range disappeared. The judgment of Glidden and Sanborn was proved: cattle became wire-wise. They did not tear themselves on the barbs as Shanghai Pierce had feared. They were confined easily within the new boundaries. Today the barbed wire bars the path of anyone who seeks to strike across the country in the West, save for some few areas so rugged or barren as to make them not worth the labor and expense of fencing.

Many Western ranches still depend on running streams or lakes for stock water. But the creak of the turning windmill wheel, the splash of the stream from the iron pipe, the ripple of the water in tanks of galvanized iron or in artificial stock ponds, and the cough of the gasoline engine when the occasional doldrums occur, are sounds now as familiar to the range as the very bawling of the herds themselves. And grass—bountifully free since the day Gregorio de Villalobos landed his first little herd in New Spain—is no longer free.

XXVIII

HIGH PROFITS AND NOBLE PROFITEERS

A gentleman who fancied himself something of a humorist and philosopher made a trip through the West in 1882 and wrote his impressions for the newspaper of which he was editor, the Lyndon, Vermont, *Union*. Later his articles were published in a curious little book entitled *The Editor's Run*, which gives an interesting insight not only into the mental workings of the editor, C. M. Chase, but of most Easterners at this period.

Editor Chase had his tongue in his cheek throughout most of his journey. He was supercilious over the gaucheries of the West; observed only the dirt and squalor of the Indian tribes; and "did not like the look of the cowboys' revolvers or the squint in their eyes."

But one thing about the West did appeal to him. When he began to talk to cattlemen and hear some of the figures on the profits of the livestock business, his Yankee eyes bulged, and he wrote home the one enthusiastic chapter in his book. His excited "figgering" follows, just as he put it down:

Your course [in cattle ranching] should be as follows: Buy a straight bunch, that is a herd of different ages. By so doing you begin to receive income at the end of the first year. If you intend to continue a long time in the business, it would be better to buy your land. This can be done by buying water frontage for such sized herds as you desire, and the range will not cost to exceed 50 cents an acre for the amount which the waterings will control. If you want to start with 2,000 head, you will aim to control 20,000 acres of land, which will cost, say, $10,000. Fence the range with wire, which will cost about $2,000 more, making $12,000 invested in the range. In purchasing a straight herd of 2,000 cows, buy cows, yearlings, and two-years-old steers. The proportion will be about as follows: 1,000 cows and two-years-old heifers, 650 yearling steers and heifers, and 350 two-years-old steers. Such a herd will cost from $14 to $16 per head, on an average. Call it $15, for the purpose of estimate, and your herd stands you $30,000. To this add the $12,000 for the ranch and the capital invested is $42,000. We will say you have made your purchase in July, when some of the cows have calves and others are coming in. But at this season the cow and calf are reckoned as one, whether the calf is born or unborn.

You are about ready to begin. But first, buy eight horses for use on the ranch, at a cost of about $400. Now brand your cattle, which will cost $100, and turn the herd into the range. One man will be the regular force. He will have a *ranch*, a mud house somewhere in the pasture, and will be required to ride past every rod of the wire fence daily, to repair breaks, and recapture cattle, if any have escaped. This service will require four horses, for he will ride rapidly and change every day. The other four horses will be kept for extra duty. Cowmen here make little account of horses, as their keeping is inexpensive, being kept on the range, near headquarters. Extra help on a cow ranch is considered equal to one-fourth of one man's time.

Some time in August the cattle are rounded up by four riders, a few hundred at a time, and the mother cows and calves are "cut out"—separated from the herd—corraled—and the large calves are branded; that is the owner's peculiar mark is burned into the hide. Then they are turned loose again with the herd. This process occupies about four days. In November the same process is repeated, and the small calves omitted in August are branded. Your mark is now on the entire herd. No extra work is required until December, when the beef buyers appear. The herd is then rounded up by a force of say eight men, who will ride two days and round up the whole pasture—get all the cattle into one bunch—and cut out the beef cattle for sale.

Now for the profits. The number of beeves sold out of a herd of 2,000 head will be about 350, and would consist of all the three and the best of the two-years-old steers. They will bring in ordinary seasons an average of $25 apiece, or a total of $8,750. The expense for the year will be $450, for the regular man at $30 a month, and an occasional helper, $125 for board, $40 for interest on the $400 for the eight horses, and $100 for horse feed and incidentals. The estimate is liberal. Total expenses, $715. Deduct this from the receipts, $8,750, leaves $8,035, or a trifle over 19 percent interest as net profit on the capital invested. Not so very remarkable after all, you say. But the story is not yet all told. Compare the size of your herd at the beginning with the size at the end of the year. The estimate is that 1,000 cows will produce 80 percent that number of calves. In order to be on the safe side, make it 75 percent, which gives you 750 calves to be added, making the herd at the end of the year 2,750 in number. From that number deduct the 350 beeves sold out, leaves 2,400 at the beginning of the second year. At $15 a head your herd of 2,000 was worth $30,000. At the same estimate your herd of 2,400 is worth $36,000. Add this $6,000 increased value of the herd to the $8,035 net receipts, gives $14,035 as the real profit of the first year, or a fraction less than 34 percent interest on the money invested.

This estimate is made on the basis that the herder has purchased his land, the purchase money being reckoned in with the cost of the herd, and so far swelling the capital investment. In free herding, which in times past has been most common, no capital was invested in land, and the profit was consequently larger. This, too, is the profit of the first year. The second year will give equal percent profit on the increased value of the herd, the third the same. It is like compound interest, every year the increase goes on drawing interest. On a five years' estimate the profit will amount to more than an average of 60 percent a year on the original investment.

The general estimate of the country is that cattle raising pays a profit, over and above all expenses, of 50 percent per annum, and that no investment can be more sure to meet expectation. I know of one case where a large investment was made in cattle a year ago, and the same cattle, with increase, growth, and rise in market, could now be sold at a net profit of nearly 100 percent. But this is owing to fortunate buying, extra grazing and a rise in market. It is an exception to the general rule.

The idyllic picture painted by the editor from Vermont is all the more impressive for his obvious efforts not to overstate the cattle bonanza he has discovered. For example, he gravely warns his reader that a net profit of one hundred percent a year "is an exception to the general rule." Quite evidently it is the utterance of a man thoroughly convinced of its truth—so much so, that its appeal to the inexperienced Easterner with capital to invest would seem to be quite irresistible. No mention is made here of the perils of blizzard and drouth, of Texas fever, of cattle rustlers, of a panic market, of the epizootic, of prairie fires, or any one of the hazards that cattlemen of experience always took into consideration. Apparently the editor had not heard of them. To him the picture was entirely rosy.

The dithyrambic writings of Chase had counterparts in other publications, some of them not so innocent in purpose. The very title of one of these indicates its intemperate tone: *The Beef Bonanza; or How to Get Rich on the Plains*. Its author was General James S. Brisbin, who set forth that he had for twelve years been a resident of the plains and gave his "impressions which shall at least have the merit of being honest"—whereat he embarked on a feverish account of the success of various "cattle kings," with testimony of governors and army officers and lush statistics to back up all the statements. General Brisbin's account, published in 1881, was as guileless a literary effort as was that of Editor Chase. The writers of both were profoundly convinced that they were revealing the secret to a virtually undiscovered treasure house. But there were other suave gentlemen not so innocent who surveyed the exciting panoramas of the West, then considered the financial centers of the East and of Europe, and conceived the notion that a promoter smart enough to combine the two would not lose by that juxtaposition.

Money was plentiful in 1881, and investors were seeking places to put their surplus cash. This was true in New York and New England. It was especially true in Great Britain. In good time sleek promoters appeared in Edinburgh and London with skillfully written prospectuses in which beautiful pictures of life on the range was limned. One such document, presented to a group of financiers in Edinburgh, told of the vast open range with free grass, described the increase of cattle, the demand for beef, and the romantic life, then went on to state modestly that the profits "had not been less than from 25 to 40 percent per annum, and under more favorable conditions would have exceeded 50 percent. Such profits are not surprising when it is kept in view that to raise a three-year-old steer worth $25 to $30 costs only from $6 to $10." The logical result of that eloquent document was the formation of the Prairie Land & Cattle Company, with a capital of one hundred thousand pounds, and a sad story of disillusionment to be unfolded as the company saw its assets crumble in the next few years through mismanagement, weather conditions and business depression.

To the younger British aristocracy the American West was a land of romance and charm, a sporting land, a land of horses and big game, and perhaps of occasional and agreeable danger. In this it differed from other frontiers, which were lands of agriculture and crude labor, and therefore less alluring to the cricket-playing, fox-hunting gentry. But the fundamental pull of the cattle country was the glittering promise of gaudy profits which was held forth.

There is little space in such a book as this to go into all the ramifications of the strange boom in cattle that occurred between 1880 and 1885—when the destruction came. Of twenty immense cattle companies in Wyoming, half in 1884 were Scottish or English, with a combined capital of about twelve million dollars. Foreign companies also established themselves throughout the Southwest, in the Texas Panhandle, and in New Mexico and Arizona. Great concerns like the Cattle Ranche & Land Company, Wyoming Cattle Ranche Company (you could usually tell they were British by that extra "e" in Ranch), Swan Land & Cattle Company, Powder River Cattle Company, and Belle Fourche Cattle Company, all were English or Scottish, and ranged in the Northwest. In the Southwest were the extensive Espuela Land & Cattle Company, the Cedar Valley Land & Cattle Company, the Clarendon Land Investment & Agency Company, and the Matador Land & Cattle Company, to mention only a few.

Money flowed west in broad channels from Boston, New York, Chicago, and Philadelphia, as well as from Great Britain, Germany, Holland and France. Cattle were on the boom. Prices rose faster than good economics justified. From Horace Plunkett, the Irish land reformer, who at the time was representing the Frontier Land & Cattle Company, came a warning when he wrote in the *Daily Drover's Journal* of Chicago, June 14, 1883: "The immense influx of cattle into the business has been to us a good deal of blind and illegitimate speculation."

Still prices mounted. Crafty old cattlemen, the few of them who could see the handwriting, began to sell out. Some British companies were cheated outrageously as they eagerly bought the cattle thus released into the market. John Clay, the rugged Scots cattleman, in his memoirs, *My Life on the Range*, recounts how the big companies, owned

by stockholders on the other side of the Atlantic and operated by hired managers, were defrauded by sharp-dealing
cattle sellers. As in all boom periods, men did business in a
large manner, carelessly. Long distance deals were made
and a custom of the country was urged upon the purchasing Britons. The West was romantic, open-hearted, the last
abode of chivalric honor, was it not? Stories had often
been told the ruddy-faced gentlemen from overseas of how
cowmen carried on large transactions by the mere nod of
the head or shake of the hand. It was natural for Englishmen to regard all Westerners as honest. And so arose the
peculiar institution known as the "book count."

The only way accurately to enumerate a herd on the
open prairie was to gather and rebrand it and keep a tally
sheet. This cost both money and time, and was often difficult to do if the country was rough or much territory was
involved. It came about, therefore, that cattle owners sometimes kept the alleged number of their cattle by book list,
adding each year the estimated increases, deducting the
sales and—ostensibly—deducting the losses. Too often,
however, the losses were not fully totaled while the increase was flagrantly overestimated. Even the banks in
this era of wild speculation were willing to lend money
freely on the security of a "book count" herd. It is not surprising then that, with inexperienced purchasers, transactions were made in which thousands of dollars changed
hands over cattle herds which were purely fictitious.

"The book count," once said John Clay, "was a gamble
in which the dealer always won. You can sometimes win at
faro, often rake in shekels at Monte Carlo, speculate on the
Stock Exchange, bet on a horse race and get your money
back and more, but a book count deal in cattle has always
proved disastrous. I have known of remnants of herds being bought and money made out of them (the parties who
have done so successfully I need not name) but on a regular book tally trade *never*."

Absentee ownership was another trouble. Old John Iliff

once enunciated the principle that a man, to be successful, had to live with his cows. He was a living exemplification of that principle to his death. But the new crop of big ranchers was not born to the sod as was John Iliff. Some were full-blooded young Englishmen from Cambridge or Oxford, and young Easterners from Harvard, Yale or Princeton, and most of them had traditions of family, wealth and ease. To be sure, not all of them were ruined by those traditions. One young cowman in the Dakotas at the time had an old family name, an inherited fortune, and a Harvard degree—but nobody called Theodore Roosevelt a tenderfoot or a weakling. He rode the range with the cow waddies, hunted bear in the Black Hills, and not only loved but learned to know passing well the cattle country.

Many young aristocrats, however, were not so energetic. The Cheyenne Club was the gathering place of the range socialites, and sitting on the piazza with a tall, cool glass, was preferable to riding the dusty flats. Then there were Denver or Kansas City or even Chicago and New York to attract the dilettante type of ranch owner. Some of these spent their winters in Italy or Southern France, and their summers on the Great Lakes, visiting their properties only twice a year, at the fall and spring roundups. Even worse was the situation of a company whose properties were owned by stockholders who lived abroad entirely, with a hired foreman or manager. The cry for dividends from the stockholders was incessant. It became, to the hectored manager, a temptation too great to withstand to sell off the herd, instead of the increase only, to pay dividends. That was like paying dividends out of the capital stock of a company, and to conceal it the manager continued to carry on his books the full number of cattle.

With the financial structures of many big companies already tottering, other things combined to undermine the solidity of the cattle industry. In 1883 railroad construction was at low ebb, bankruptcies were mounting, many businesses were passing dividends, and receiverships were increasing. Still the cattle boom grew in the West. For a while it appeared that cattle would hold up, independent of the rest of the business world.

H. L. Goodall, veteran editor of the Chicago *Daily Drover's Journal*, sought to put a brake on the steady stream of range cattle being shipped to the big markets at Kansas City and Chicago. He pointed out that foreign competition had arisen, with Russia and Germany building packing plants to cut down the business that hitherto had gone to the big American packers. Yet the cattle market only staggered higher. Nothing, it seemed, could cool the fever of speculation, and the eager buyers expected prices to go to $40 or even $50 a head. But in the fall of 1884 there was an ominous occurrence. Through the Chicago Livestock Exchange went a report that ranges were overstocked, Texas fever was rampant, and the open range was being reduced by barbed wire. The price of range stock dropped to $2.50 a hundredweight. Beef cattle did not fall so low, but the beef market was affected. And Goodall punned in his *Daily Drover's Journal*: "If you have any steers to shed, prepare to shed them now."

Yet even then there was a rally. The year 1885 was the last of the cattle boom. In May of that year fat cattle brought $4.75 a hundredweight, but by fall the price had begun to slide. In October fat cattle were bringing as little as $3.00 and stock cattle had fallen as low as $1.80 a hundredweight. And then, as if to add the final cap-sheaf to the woes of cattle land, Nature struck a stunning, sickening blow.

ICE-SHEATHED DEATH

Never was a softer, more pleasant autumn than that of
1885. On the high plains the cattle grazed and grew fat in
days that were one long succession of mildness, with a soft
Indian summer haze always on the horizon, making the distant landscape seem phantom-like. The nights were pleasant, too, and a man could roll up anywhere in his blankets
without feeling chilled before dawn.

Among cattlemen optimism was general. The market, it
is true, was bad, but there were signs which most took to
mean that prices might rally as they had always rallied before. Meantime the Northwest was proving a peerless cattle range. The cattlemen congratulated themselves on their
discovery. Perhaps winter weather was sometimes severe in
the Northwest; but in the main its rigors apparently had
been vastly overrated. Since the introduction of cattle on
a large scale in Wyoming, Montana and the Dakotas, there
had been nothing to threaten any general losses. A little
winter killing, to be sure. But the winter kill was to be ex-

pected and at least there were no mammoth die-ups here from drouth as sometimes occurred in the Southwest. What if the market was unfavorable? So long as the range remained superb and the weather a proven factor, the herds would continue to increase and the cattlemen could afford to wait.

That was the general feeling. Those cattlemen who had planned winter journeys departed for New Orleans or Europe with unworried minds. A general comfortable, satisfied buoyancy imbued the range country.

The last day of December, 1885, dawned clear and mild, with a low barometer and a peculiar yellowish purple bordering the northern horizon. Before noon in Nebraska and Kansas a drizzle began, growing cold as the temperature dropped rapidly. Presently the rain turned to snow and the wet ground froze solid. The snow thickened, became a fine white powder of razor-sharp crystals, a choking smother, swept along by a hurricane wind. By evening the thermometer registered below zero and was still tobogganing. And by night the range country from the Dakotas south into Texas was lashed by the greatest blizzard in history, which hid all landmarks, obliterated all trails, and brought a deadening cold with it such as men in the area had never known before.

And the cattle? They had been bred for no such clime as this. From the South they had come and many of them were scarcely yet acclimatized. As the knife-edge gale swept level across the plain with its choking burden of ice crystals, the cattle drifted with it. To drift was their only chance. It was impossible for man or beast to breathe without turning the face down wind, for the blast seemed to snatch the air from the lungs. The fury of the wind was heard in the insanity of its howling.

Throughout the cattle country men shivered and piled another chunk of wood in their stoves. The hope was that the blizzard would soon pass. But there was no abatement the second day and by the following night the situation was alarming. Not once had the terrifying wind diminished. Every object as much as fifty feet away was invisible because of the curtain of fine driving snow.

Here and there, through the West, and in growing numbers as you went eastward, were the homes of nesters—the little farmers who had taken claims or established homes by squatter right in the open country. In a storm where there was almost no chance for cattle and horses, there was little chance for men unless they were warmly housed and protected. Most homesteaders lived in shacks which were makeshifts at best. They could afford nothing better—and besides a claim was usually for quick sale, the building of permanent improvements being regarded as a waste of money. An average shanty of the day was about ten feet by twelve, built of cheap lumber, and heated by a sheet-iron stove. Entire families were trapped in such huts. Accounts of the day tell of such incidents as this: A man, returning from a town in which he had been when the storm struck, found his flimsy home blown down and his family, seven in all, scattered about in the snow, frozen to death. At another place, three men were found frozen and dead in a wagon, with frozen, dead horses. Although they had huddled together for warmth the cold had killed them.

The dugout, most primitive of Western dwellings, proved the best refuge in the great blizzard of January, 1886. Families in dugouts generally pulled through. Those who lived in the thin-walled shacks too often froze in their homes, and such as were saved by rescuers were generally found to have taken to bed, covered themselves over with every blanket and comfort available, and waited there for succor.

All the stock on the ranges in the Dakotas, Nebraska, Kansas, the Indian Territory and the Texas Panhandle was on the move, while parts of Wyoming and Colorado were affected. A dispatch in the Dodge City *Daily Globe*, January 11, said:

The heavy snow and bitter north winds of the past ten days have caused the most serious apprehension among cattlemen as to their probable losses. Up to this time but few have come in from the range country, but within a few miles of here no less than five hundred head have drifted to the river, where they perished in attempting to cross, or drifted up to fences, where they remained until frozen to death. A gentleman from a ranch south of here reports seeing cattle on his way up that were still standing on their feet, frozen to death. The water holes are frozen over, the grass is snowed under and the weather is cold, with every prospect of more snow. The loss of livestock is bound to be very heavy on the Arkansas River, as cattle are drifting down from the Kansas Pacific road.

And from north of the Kansas Pacific, too. On the third day of the storm many cattle outfits sought to follow their herds. Cowboys dressed as warmly as they could and plunged their horses out into the snow, swearing and whooping, their chuckwagons following along behind them. How many of them died nobody will ever know. The Kansas Historical Society has a list of more than one hundred persons killed in that state alone by the blizzard of 1886, and many of these were cowboys. Some of those who fell were picked up by the heroic ranch cooks in their chuckwagons, but by the time a cowboy was so far gone as to fall from his saddle, there was little hope that anything could be done for him.

For the cattle it was simply murder. Doc Barton had nearly twelve thousand head under his brand, the O S Bar —eleven thousand grade cattle and eight hundred registered animals. He lost in the storm every head of his registered stock and nearly all his grades—and what survived he rounded up as far south as Texas where they had drifted with the storm. R. K. Farnsworth's Circle M had six thousand cattle before the storm; afterward one hundred and eighty-two were found. Of fifty-five hundred cattle on another Western Kansas ranch, five thousand perished at a loss of more than $100,000. Another rancher was offered

$25,000 for his cattle a few days before the storm and refused. After the blizzard he sold the remnant for $500. Multiply these instances by the whole extent of the tier of states over which the storm swept, and you have an approximation of what that blizzard of 1886 meant to the cattle industry.

Never before or since has there been a loss of livestock so unprecedented. Hundreds of outfits were completely wiped out. It was the greatest die-up in the history of the West, and it prostrated the whole area affected.

Skinners, as after every blizzard in the West, reaped some kind of a harvest by peeling hides from the carcasses after they thawed; but after that the bloated bodies were left to decay or be eaten by coyotes and crows which grew fat. Buffalo Jones wrote of what he saw:

As I drove over the prairies from Kansas into Texas I saw thousands upon thousands of carcasses of domestic cattle which had drifted before the chilling, freezing norther. Every one of them had died with its tail to the blizzard, never having stopped except at its last breath, then fell dead in its tracks.

But Nature was not through. The summer of 1886 was dry, with little water and correspondingly poor grass in Montana, Wyoming and Colorado, so that the cattle which escaped the blizzard of January 1886—which missed the western area of the Northwest range country—came up to that autumn thin and in poor condition. This time the winter set in early. The first snow fell in October—a snow so heavy that it blocked some of the smaller streams. After that there never was a time when the snow left the ground until the succeeding spring. The cattle, unable to reach the grass, grew skeleton thin and weak.

Worse was in store. In December came a succession of blizzards. None of them was as severe as the great blizzard of the previous January, but in the aggregate they were almost as deadly and they almost exactly covered the coun-

try which had escaped the great storm of the winter before. One followed another, dealing blow after blow at the
weakened herds. Then in January, 1887, came a chinook—
a soft, warm wind which partly melted the snow into a
thick, mushy slush. It was followed by a hard freeze and
more blizzards. That chinook and its succeeding freeze
turned the whole Western plains area into a solid chunk of
flinty ice. There was no question now of pawing down to
the grass—even for horses. And still blizzards swept down,
one after another. At times the temperatures reached forty
degrees below zero.

When spring belatedly came the ranchers hardly needed
to bother to count. They knew themselves wiped out as
completely as had been the cattlemen of the more easterly
regions the previous year. Canadian and American ranchers
alike were bankrupt. Struthers Burt has described that
roundup of the spring of 1887 as the quietest on record.
No joking, singing, or horse-play. Just the silent, strained
look as men counted in the coulees and along the streams
the heaped carcasses of their cattle.

Nobody ever enumerated the number of cattle killed in
the two blizzard years of 1886 and 1887, but few indus-
tries in the history of the world ever before felt a combination of blows so crushing. Eighty percent of the cattle in
large sections of Wyoming were killed and other states lost
in proportion. Ike Pryor, who was grazing a big herd in
Colorado, was almost completely wiped out. Years later
John Clay wrote in his memoirs:

It was not till the spring roundups that the real truth was
discovered and then it was only mentioned in a whisper.
Bobby Robinson, acute judge of conditions, estimated the loss
among through cattle at less than fifty percent. It turned
out to be a total loss among this class of cattle and the
wintered herds suffered from thirty to sixty percent. . . . It
was simply appalling and the cowmen could not realize their
position. From Southern Colorado to the Canadian line, from
the 100th Meridian almost to the Pacific slope it was a catastro-

phe which the cowmen of today who did not go through it can never understand. Three great streams of ill-luck, mismanagement, and greed met together. In other words, recklessness, want of foresight, and the weather which no man can control.

Great firms like the Continental Land & Cattle Company, which ran thirty thousand cattle before the blizzard winter were reduced almost to nothing. The Worsham Cattle Company, a British firm, simply quit—made no attempt at the tragic effort of gathering up its pitiful remnants. The Swan Land & Cattle Company, a Scottish firm, which had fifty-five hundred three-year-old steers on the range the previous fall, ready to market, found about one hundred the following spring, and their stock cattle were reduced in proportion. Swan, Sturgis and other great companies were ruined; Kohrs, Murphy, Granville Stuart, Joe Scott, and others badly crippled. Among those who disappeared from the range after the disaster of 1886 and 1887 was Theodore Roosevelt, who left the bones of his cattle behind him, north of Medora.

A disaster so complete had never before been felt. Not even the wholesale thefts of the Comanches in the old Texas days could compare to the white smother that wiped out the entire industry. Many British ranchers went home in anguish of spirit, never to return. But others remained, fought through the long hard years, and gave a priceless contribution to the land of their adoption. Such men as Murdo MacKenzie, W. J. Tod, John Clay, and their fellows refused to be defeated by misfortune. They stayed to help rebuild the shattered West.

But the era of great cattle companies was gone, along with the wild extravagance, the toplofty ideas, overstocking, frenzied financing, mismanagement and the rest—to a large degree at least. These things never can be wholly eliminated, but their great flood tide was past. The day of untempered speculation was at an end. Prostrate, the cattle country after the catastrophe was at least free to start on a sounder basis for rebuilding. As one old cattleman said, "The water wasn't squeezed out of the cattle business, it was froze out."

Long years were needed to come back, but the courage of the West never had a finer demonstration than in this gloomy period. Neighbors helped neighbors. Men with a little money came to the rescue of men who had none. Ike Pryor was saved by a commission house to which for years he had consigned his cattle. Major Andrew Drumm financed dozens of cowmen. Even the packers, perceiving the peril in which the industry stood, came to the rescue with loans.

When, after a stern and lengthy struggle, the cattle country once more stood on a sound basis, it was a changed and chastened region. The fence had come and with it the conservative banker. Stock raising had in some measure ceased to be a wild and glorious gamble and had become a sober business.

6. . . . BLOOD ON THE SADDLE

XXX

COLONEL COLT'S EQUALIZER

Sam Colt had an inborn instinct for guns and a hankering to whittle. The last was natural for he was a Yankee, and you can observe the same craving to do something with restless hands in a group of village loafers on the steps of any New England crossroads store today. The instinct for guns was natural too. It came from a long line of fighting forebears, including his grandfather, old Major John Caldwell, who had carried a sword in the Revolution, told hair-raising stories concerning the exploits of Tim Murphy, and gave the boy an old horse pistol for a keepsake.

Tim Murphy was a legendary scout of the Iroquois country, who performed prodigies against the British and Indians with a double-barreled rifle, and his adventures stimulated young Sam Colt to wonder whether the principle of multiple shooting might not be carried farther even than two shots to a gun. By the time he was sixteen and sailing aboard the brig *Corlo*, India-bound from his home in Hartford, Connecticut, a seaman before the mast, he caught his great idea from watching the ship's steering wheel; noting the manner in which, however the wheel turned, each spoke fitted directly into line with a clutch which held it fast. That was where the whittling knack came in handy. In his off-watch leisure, Sam Colt whittled out the model for the first pistol with a revolving cylinder, which locked for each shot and was operated automatically by cocking the hammer. It was the idea behind the Colt

patent and was destined to revolutionize the world's firearms industry.

The long fight of the young inventor to obtain financial
backing, and then to induce the government to give him a
hearing, makes an interesting story but has no place here.
While he was striving desperately and fruitlessly to induce
the United States army to depart from its conservatism
long enough to adopt his vastly improved repeating revolvers and carbines, a market was open to him; a market
he discovered by pure accident.

Down in Texas the new young republic was fighting for
its life against the Mexicans on one side and the Comanches
on the other. Credit is generally given to S. M. Swenson,
merchant and cattleman, whose S M S brand (the S's re-
versed) is still being run in Texas, for introducing Colt
revolvers into the cattle country. Through him—or somebody else—a few of the repeating pistols reached San Antonio sometime before 1839 and were turned over to Captain Jack Hays and his famous company of rangers. For
years the horsemen of the plains had needed a weapon exactly like this—a gun to be carried in a holster at the side,
capable of being quickly drawn, and of being fired several
times without reloading. The rangers took the Colt revolvers to their hearts.

A few months later, Samuel Colt, invited to the New
York offices of Samuel Hall, the city's leading gunsmith,
was introduced to a tall, tanned man who bore the name
of Captain Sam Walker. This was the noted Texas ranger.
They were three famous Sams who gathered in that room—
Sam Colt, Sam Hall, and Sam Walker—and the circumstance must have been propitious, for it was a historic series
of events that started from that meeting. Captain Walker
had been sent all the way east to obtain some more of
Colt's revolving pistols for his border fighting men. Walker
had some ideas of his own. At his suggestion the "Walker
model" was built by Colt, embodying a heavier and
stronger frame, a grip of more convenient shape that fitted

more naturally to the hand, and an improved loading device. Eli Whitney, the inventor and manufacturer of the cotton gin, made the first consignment of revolvers ordered by Walker for Texas, on specifications furnished by Colt.

From the beginning the cowboys as well as the rangers of Texas had a beautiful affinity for the six-shooter as it came to be almost universally known. Sometimes it was also called six-gun, hog-leg, shootin'-iron, hardware, artillery, and other names, all given in affection, because the revolver was perfectly adapted to the Westerner's needs. The Texan-Sante Fé filibustering expedition of 1841 was equipped with Colt rifles and revolvers, and one of its battles with the Kiowa Indians is the first written record of the use of Colt guns in actual fighting. In 1846 when the Mexican War broke out, Jack Hays organized a regiment of Texas rangers, armed them with Colts, and marched at their head as colonel with Zachary Taylor's army to Mexico where they won for themselves a bloody renown.

After that demonstration by the Texans, the army adopted the Colt. But by then it was in almost universal use among cattlemen. So much a part of the standard equipment of every rider of the range did it become, that until comparatively recently, a cowboy in the West felt that he was scarcely decently garbed when he appeared without his cartridge belt and the six-shooter in its holster.

Events were shaping themselves to make the Colt revolver a factor of increasing importance in cattle land. When Charles Goodnight trailed his first herd of longhorns into the Palo Duro canyon, he started a movement by which the Texas Panhandle became overnight the domain of the nation's greatest ranches, thus occupying the last unsettled portion of the West. Within a little more than a decade the tidal wave of cattle had swept out from Texas, and with the final occupation of the Staked Plains, civilization—or its precursor—had taken over a territory comparable in size to that which required three centuries to win east of the Mississippi.

Grass and water had been free for all. A man had only to point his cattle toward the outer rim of the range country to be assured of good grazing land on which to establish a ranch, although to be sure he often still had to fight for it. But before many years the uttermost limits of possible cattle country were reached. Every valley and plain was explored. Every water site was taken.

The movement once started was, however, difficult to stop. With a sort of inertia the herds continued to roll outward from Texas. Information traveled slowly and men were too occupied on the far frontiers with matters of present moment, such as Indians and blizzards, to send word back to the folks at home that the range was full. The West, moreover, had an adage: "Let every man kill his own snakes." It was predicated on the general unwillingness of the frontier to try to tell another man his business.

Down in Texas, therefore, trail drivers continued to form herds and take them north or northwest or west, perfectly oblivious of the fact that they were entering territories already very much occupied. The herds began to flow into one another and elbow room grew scarce in New Mexico, Arizona, Wyoming, Montana, the Indian Territory—everywhere. To complicate this situation came the barbed wire fence. The big outfits were putting wire around the holdings they claimed. Water, a prime necessity everywhere, began to be denied the weaker and smaller ranchers. The nesters were coming in and the man of the plow took yearly more of the land from the man of the saddle.

A thousand tensions were ready to burst into violent action. Feuds broke out—feuds between ranch and ranch, between big operators and little ones, between cattle owners and cattle thieves, between stockmen and grangers, and between cattlemen and sheepmen. They were to write bloody history over a territory as large as Europe, within a period starting about ten years after the beginning of the trail drives and continuing up into the present century. Those feuds were symptomatic of the thing that ailed the range country—overcrowding.

The Colt revolver played a big role in this turbulent period of the West. It barked its staccato messages of death in the saloons and dives of the trail towns—Abilene, Newton, Dodge City and the rest. It did its share in vanquishing the Indian. But it was in the wars of the lean, sunbaked riders of the range, and in the eventual bringing of law to the cow country that it played its most decisive role.

Billy the Kid was a slight youth, inches under six feet tall, with narrow, sloping shoulders, and small hands. Physically he was no match for any able-bodied man of normal size and strength. But Billy the Kid held in terror half of New Mexico and part of Texas. That was because of his lethal virtuosity with the nimble weapon he swung at his belt.

Doc Holliday was a skeletonic consumptive, but his name brought respect from men far bigger, healthier, and fully as brave as he. It was because of the holstered death at his side.

Countless other men in the West lacked the physical equipment of their associates, yet were able to establish themselves on a high plane as fighting men by their skill with the revolver. Bullies were scarce in the cow country. It was because all men were measured by the standard of the six-shooter—and a big man died just as quickly from "lead poisoning" as a little man.

In the West they called the six-shooter "Colonel Colt's Equalizer."

XXXI

TWILIGHT OF THE LONGHORN

Unquestionably the biggest single revolution in the cattle country was the introduction of well-bred sires into the longhorn herds. Of a necessity this followed the coming of barbed wire fences, and, to a degree, of windmills, water wells, and the nearer markets which the railroads made available. Purebred cattle, whether Shorthorn, Hereford, Angus or any other strain, were not nearly so hardy as their longhorn brothers, nor would they survive under conditions in which a longhorn would do fairly well.

As early as the '70's attempts were made in Texas to grade up the range herds. Goodnight's experiments with Shorthorns have been mentioned. There is a question whether William Powell or W. S. Ikard first introduced Herefords—the hardy, red-bodied, white-faced cattle which are so predominant now in the West. Ikard attended the Centennial in Philadelphia in 1876, saw the Herefords exhibited there, and conceived the notion they might make good range breeding stock. Both he and Powell brought Herefords into Texas in 1878.

Other ranchers gradually followed their example. Not, however, until the '80's did there begin to be any real change in the appearance of range cattle. Except for a few white-faces or Shorthorns here and there, the herds remained the same—long-legged, long-backed, long-horned creatures, variegated in color, wild as deer, stringy, lean and vicious. They had only one peerless quality—they could live under conditions where softer breeds would perish.

Introduction of purebred sires was expensive. As an example, O. H. Nelson in 1882 drove a bunch of Herefords into the Panhandle. The herd contained twenty-five registered bulls which cost $250 a head, and six hundred cows with four hundred calves at foot. Goodnight bought this entire herd for his J A ranch, taking bulls, cows, and calves, at an average of seventy-five dollars a head—several times the prevailing rates for Texas cattle, then bringing about fifteen dollars a head.

Goodnight herded the Herefords separately and improved a select bunch of stock which he gave a special brand—J J—a brand long considered the blue-blood insignia in the Panhandle. Earlier, Lee & Reynolds, who had started as sutlers at Fort Elliott and then gone into the livestock business, bought seven carloads of Herefords for their ranch, and other Texans also began to import whiteface bulls.

In the late '70's and early '80's the Shorthorn was by all odds the most popular beef animal in America, partly because it was one of the oldest recognized breeds on the continent, and partly because it was blocky, milked well, and had a prepotency which brought its crossbred get well up toward the size of the purebred cattle. A crossbred Shorthorn calf would dress out three hundred pounds heavier at two years than a plain longhorn dogie. Not a few cattlemen experimented with Shorthorns, but circumstances prevented the full success of the breed in the West. For one thing the best sires were in such demand in the East that only the culls—classified as "range bulls"—were available for shipping across the Mississippi. Because they were receiving inferior cattle of the strain, Western cattlemen developed a prejudice against the whole breed of Shorthorns, considering them unthrifty.

Throughout the West, meantime, the Hereford was winning popularity. The red hide, white face, white lineback, and white underparts became familiar. The Hereford has a striking ability to reproduce its color and markings, and a bunch of white-face calves of the same age look as alike as so many peas in a pod. This fine uniform appearance added to the breed's popularity. Of recent years, however, cattlemen have begun to suspect that they can obtain excellent results by crossing Shorthorns with white-face stock, to make a bigger, blockier, beef animal and adding finish value.

None of this, of course, was accomplished in the old free range days. Even if a man desired to grade up his cattle he could hardly be expected to be so philanthropic as to invest his money in expensive registered bulls which would, in that unfenced area, produce as many calves for his neighbors as for himself. Returning to the Nelson bulls: they were bought by Goodnight at $250 a head. One bull to each twenty-five cows was the common ratio, and the investment for a herd of any size at the rates Goodnight was paying was enormous. The cost of buying enough bulls for a herd of one thousand cows was $10,000—and the entire herd of range cows was not worth much more than that.

But improving the breed was the new order of the cattle industry. The fine, "slick-haired" stock rapidly began to supplant the old, dour longhorns. One of the most arresting changes in the West has been the appearance of its cattle. No longer is the herd a mixture of colors, of long, curved horns, and of wildness. As alike in size and markings as a uniformed army it is now, and with a docility that removes it far from the adventurous trail herds of the old days. The true longhorn has almost disappeared. Here and there an occasional survivor can be found, kept as a curiosity. But the breed is more scarce in the West than the buffalo it supplanted.

When he came into competition with one of the powerful big cattle companies in fencing projects, the little cattleman generally suffered. Some of the large concerns at first leased their land from the state. Later considerable areas were obtained through buying up scrip from ex-soldiers, who obtained it as a bonus for their war services.

And it was no rare thing for a cattle baron to fence in land to which he had not the remotest legal claim.

Very frequently the big companies used unscrupulous means to force out smaller neighbors. A little rancher might be "fenced in"—surrounded completely by wire through which he could obtain neither egress nor ingress,

so that he was forced to sell out at a figure named by his freebooting neighbor. Ill-feeling stimulated by such tactics caused various local disturbances known as "fence cutters' wars." Sometimes masked, nesters and small ranchers banded into secret organizations, and moved about by night snipping wire strands. The cattlemen retaliated. Houses were burned, crops destroyed, men even killed on one side or the other.

In the fall of 1884 some small land owners in Custer County, Nebraska, charged that the Brighton Ranch Company was fencing in their holdings and demanded that a stretch of fence inclosing a pasture fifteen miles square be torn down. No attention was paid to this demand by the ranch company. The next step of the landowners was to cut the wires and carry off the posts of the objectionable fence to be used for rafters in settlers' sod houses.

The foreman of the Brighton outfit promptly swore out warrants for the fence raiders, secured their arrest, and saw them taken off to the county seat, Broken Bow, for trial. Then, while the nesters were gone, the foreman with his cowboys visited every one of the "soddies" in which the fence posts had been utilized, and removed the logs in a forthright manner. A team was hitched by a chain to the end of a rafter, and out it was jerked. The sod house collapsed—a heap of dirt, hay, roots and debris, covering the furniture, utensils and other pitiful property of the homesteader.

While the cowboys whooped and skylarked at this congenial destruction, a boy who had witnessed it rode hard for Broken Bow with the news. Immediately a posse of a hundred angry men leaped into their saddles and started for the scene, but the Brighton riders had been warned and were gone when the Broken Bow citizens arrived. Public opinion so reacted against the Brighton ranch that the homesteaders who had been arrested for destroying the fence were released, and the Brighton foreman was arrested, tried and fined for tearing down the houses of the citizens.

In most wire-cutters' wars, however, the homesteaders were not so successful, for very often the big cattle interests controlled the courts and county governments. The day of free grass was over, but many in the West still held to the old tradition that a man should have a right to graze his cattle as far as they could walk from his water. Because the great companies could afford to put the most land under wire, it seemed at first that the new trend was putting all the grazing lands into the hands of the big interests. Actually it worked out in the end the other way—the big ranches were practically driven out of existence. But nobody for the present foresaw the slow infiltration of small nesters, the gradual breaking up, here and there, of big ranch properties, the devastating effects of depressions, panics, drouths and blizzards.

Because these things were not foreseen, public sympathy, always leaning to the underdog, became strong for the small ranchers and had its ramifications in many directions. Before the War between the States, cattle rustling had been uncommon in Texas, where an almost chivalric sense of honor prevailed. Charles Goodnight told J. Evetts Haley once that before the War it was an unwritten law, almost never violated, that every cattleman should mark and brand each calf found on his range to its proper owner, and send strays as far back toward their own range as possible. Double care was taken in branding cattle belonging to widows or orphans. Sometimes cattlemen actually leaned backward a little and branded for such unfortunates mavericks on which the title might have been a little difficult to prove.

During the War, while most of the able-bodied Texas men were in the Confederate army or with the rangers, cowboys remaining on the range were so few that two or three calves were left unmarked each year for every one branded. A growing crop of mavericks was the result—and a growing temptation that some men could not resist. Surreptitiously a few individuals began to brand maverick cattle over to themselves. The practice resulted in some feuds and brief, flaring fights, and eventually it was stamped out as far as open mavericking was concerned. But the rustlers continued furtively to prey. To this day they remain a constant jackal-like menace to the cattle industry.

When barbed wire first went in, resentment against the big owners took the form of public approval of everyone who opposed them. The small cattle thief for the first time found himself bracketed in the public mind with the honest small cattleman—since the big operators were their common enemies. The idea seemed to be that anyone who damaged the intolerant overlords of the range was acting for the general weal. But it did not take long for the thieves to go past the limitations of the vicarious cloak of respectability which this new public sanction gave them.

At first there were many border line cases. In Rice County, Kansas, one cold, hard winter, there was a big die-up. Thousands of cattle perished and of the survivors many, huddling along the creek bottoms, were almost dead. Some money was made by the small settlers through skinning the dead animals for their hides. The skinner usually received two or three dollars a skin, which he split with the owner of the brand on the hide. This was legitimate and well enough, until it suddenly occurred to the big cattle companies that perhaps they were losing cattle which might have survived. Coming on cows that appeared to be dying, settlers sometimes shot them to "put them out of their misery." Often, the owners strenuously declared, the cattle that received this *coup de grâce* had a good chance to survive.

A bloodthirsty announcement was made: the first settler found skinning a cow would be shot on sight. It was a declaration of war, because there still remained many unskinned cattle and the revenue was badly needed by the nesters. The skinners held a meeting and issued a reply to the ranchers' ultimatum. If any of their number was killed by the cattlemen, the prairie would be set afire. The result was a stalemate. Neither side dared move; the nesters for fear of death, the ranchers because they would lose all their remaining cattle by starvation if the grass were burned off.

From skinning dead cattle, to shooting cattle that might perhaps be dying, finally to stealing living cattle, was no long series of steps. Rustling became, in effect, a protest against the illegal holding of the range by the big companies. Of course the dispute served many persons who were not legitimately interested in either side of the argument and merely wanted to profit dishonestly. Nor should it be forgotten that the charge of rustling was a pretext used on occasion by the big ranchers to drive honest settlers out of the country. As the situation grew more tense, big rewards were offered for information leading to the conviction of cattle thieves, and when convictions occurred, shrift frequently was short.

Two settlers named Ketchum and Mitchell lived together in a sod shanty near Broken Bow, Nebraska. A posse surrounded their cabin and ordered them to surrender on a charge of cattle stealing. Ketchum and Mitchell resisted instead, and in the brisk little fight that ensued, Robert Olive, a ranchman, was killed. Olive's brother at once took command of the posse and began a grim siege of the cabin. Eventually Ketchum and Mitchell exhausted their ammunition and surrendered. At once Olive's cowboys hanged the two men to the nearest cottonwood.

When the lifeless bodies were discovered and cut down the rancher and his men were arrested. Local indignation was great, so the trial was held in another county in justice to the prisoners, all of whom were convicted and sentenced to life imprisonment. They appealed to the state Supreme Court on the very issue of the attempt made to give them full justice, and the Supreme Court upheld their contention that they were entitled to a trial in their own county. On the day the case was to be heard in Broken Bow, stern-visaged horsemen rode into town, their six-shooters suggestively prominent. In the courtroom those same hard-eyed men sat and watched the proceedings wordlessly. Every person who entered the room encountered the calculated stare from a score of grim faces. No witnesses appeared against the defendants; it was concededly unhealthy to do so that day. For lack of prosecution all the prisoners were released.

A vigilante posse was extremely unhappy one day in the '70's. A perfectly good prisoner, a length of raw-hide lariat, and a cottonwood tree, the widespreading branches

of which seemed especially designed by a beneficent Provi-
dence for an impromptu gallows, would have seemed
enough to satisfy any self-respecting group of stranglers,
out from Dodge City to put a stop to cattle rustling. But
it wasn't.

The truth is that the vigilantes were in the painful process of surrendering a cherished hope. They had set their
hearts on a hanging and now they found that even by
frontier custom their desire could not be carried out.
Hence the men stood around in moody groups, spat dis-
consolately, and swore dolefully. And nobody who knew
the situation could have blamed them much.

The prisoner, sitting between his guards, was a Negro—
Ben Hodges of Dodge City. Rumor had accused him of
numerous crimes, including cattle and horse rustling. Two
days before a herd of steers had disappeared from a corral
near Dodge. The vigilantes, feeling sure they knew who
had caused the disappearance, set out to find Ben Hodges.
They traced him to a remote canyon and it appeared that
a neck stretching had arrived for him at last. They made a
surround, charged his camp, and before he could reach
for a weapon, the Negro was a prisoner. The trial was
brief. Ben was notified he would be strung up and everything was ready for the ceremony, when some killjoy
asked a question. "Where are the stolen cattle?"

The silence that followed was tomblike. Not a man
present but knew the hanging was definitely, perhaps permanently postponed. Every man there was morally certain
that Ben Hodges had stolen the cattle. But there was no
actual proof of guilt. Some there were who hopefully
urged the hanging "on general principles," but the suggestion was half-hearted. Then someone had an idea, ex-
pounded it, and in spite of the horrified screams of protest
by the prisoner, the posse carried it out. In lieu of hanging Ben Hodges, they simply crippled him for life, by
severing with a knife the tendons of both heels. It was
cruel—but it was frontier justice. Hodges had been spared

his life. It seemed to all of the vigilantes he had escaped easily.

Curiously, the Negro outlived every man of the posse that crippled him. Until a few years ago he occupied a dilapidated hovel on the outskirts of Dodge City, with the mark of his terrible experience with the "stranglers" ever noticeable about him. As a boy in Dodge City, I knew him and talked to him frequently. In addition to the natural infirmities of old age, Ben was handicapped by being unable to walk except at the laborious shuffle to which he had been condemned by the slashing of his heel tendons. About his house and garden in Dodge he worked on his hands and knees. He was a well-known figure in the old cowboy capital as late as 1926, when he died, puttering about his garden or dragging his springless legs down the street to the last, a comical grin on his wrinkled black face.

Hodges never seemed to hold resentment against the men who so frightfully maimed him. He originally came to Dodge City as a cook in one of the first trail herds. Half Negro, half Mexican, he was endowed with quick natural intuition and an active imagination which enabled him to inveigle not only tenderfeet, but even super-cautious bankers and lawyers into believing that through his Mexican mother he had a claim on large Spanish land grants in New Mexico. He obtained letters of credit and bankers' endorsements, but although he was able through them to borrow money he never could repay, he was never able to establish his claims.

Frequently the Negro was in trouble prior to the heel-cutting episode. Horse Thief canyon, seven miles from Dodge City, was the rendezvous of a gang of stock thieves, and Hodges was accused of being "custodian" of the canyon and of keeping camp for the gang. This was never proved, but his reputation as a bad actor grew. More than once he faced a jury, on one occasion escaping a possible death sentence on a murder charge when he harangued the jury with such absurd and imaginary tales that they pronounced him "not guilty" in sheer gratitude for the way he had diverted them. Old-timers at Dodge City said Ben's chief activity was hiding stock in Horse Thief canyon until a reward was offered, then suddenly "finding" the animals and claiming the reward. The day the Horse Thief canyon rustlers were captured and seven of them hanged at once on a single huge cottonwood tree, Ben was not present. It was shortly afterward that his ankle tendons were slit.

When Ben Hodges died he was more than one hundred years old, and claimed to be the oldest citizen of Dodge City. Notables came from all over the country to attend his funeral. He was a symbol of a lawless period, and a living example of a justice which was sure if sometimes horrible.

XXXII

THE GRIEVANCE OF JOHN CHISUM

When John Chisum, after some seasons of trailing herds over the Horsehead route for Charles Goodnight and others, decided to go into ranching for himself in New Mexico, he was making no novice's plunge. A man born to large affairs in the cattle country was Chisum. Even before the War between the States he made cattle drives to Shreveport on the Red River, whence shipments were made by steamboat to Memphis, Vicksburg, Natchez and New Orleans. He founded Paris, Texas. And he had behind him a varied and somewhat spotted career in business and finance.

After the War, Chisum made one venture in the packing business. He entered a firm, under the name Clark, Chisum & Wilber, to operate a packing house at Fort Smith, Arkansas. It is said that part of the money for this venture was furnished by S. K. Fowler of the British packing family. Chisum's function in the firm was to buy the cattle on the Texas plains and deliver them at the plant. His partners were supposed to run the establishment. But the process of packing in those days was crude, the meat spoiled, and the concern went bankrupt. Convinced that his forte was not the packing business, Chisum made the alliance with Goodnight to trail herds across to the Pecos.

It is probable that Chisum believed he was leaving behind him the entire episode of the ill-starred packing venture, but in this he was in error. Notes had been given for some of the cattle purchased for the plant and of these there were some $80,000 to $90,000 worth outstanding, with Chisum's name on them along with other members of the company. Creditors of the firm, finding that alone of its active members Chisum had not taken bankruptcy, sued him and obtained judgment against him. This made John Chisum laugh. He claimed, first of all, that in spite of the use of his name by the firm he was not a partner, and had been assured legally there was no just claim against him. Secondly, he asserted that he had lived more than four years in Texas after the failure of the company and that therefore the statute of limitations had run on the obligations. Finally, he owned nothing but a lot of wild Texas steers, and there was no market for them.

So he turned his face west and began driving cattle across the Staked Plains. His surplus unsold cattle he located at the Bosque Grande on the Pecos about thirty miles north of the present city of Roswell, where he built his first ranch headquarters. And there the Long Rail and Jingle Bob had its beginning. In his first years at the Bosque Grande he continued to drive cattle from Texas. Gathering herds on the open range was sometimes difficult. To simplify his task, Chisum obtained from the various cattle owners in the districts he was working powers of attorney that permitted him to pick up cattle of their brands, incorporate them in his herds, drive to New Mexico, and presumably pay for them after delivery. It was charged later that he overlooked in some cases the payment process.

Meantime the Jingle Bob holdings grew rapidly. Up and down the Pecos ranged Chisum's cattle for one hundred and fifty miles and as far on either side as "a long-legged longhorn steer could graze." The owner of the Jingle Bob was called the King of the Pecos and his herds at one time were said to number one hundred thousand head—more than any other man or outfit held at the period. After a time Chisum moved from the Bosque Grande to South Spring River, forty miles down the Pecos. Here a never-failing spring of great volume gushed from the earth, and here over a period of years he built his famous South Spring ranch house, a room or a wing at a time, until it was one of the authentic palaces of cattle royalty, and in its day a show place of the Southwest. It was a great, rambling adobe structure, plentifully varandaed, plentifully furnished with bunk houses, one story high after the Spanish manner, and surrounded by fruit trees, cottonwoods,

and alfalfa fields which flourished under irrigation. Here John Chisum dispensed a right royal hospitality. Rare was the day when his long table was not filled with guests, for in ancient baronial fashion all who passed in that wild country were urged to stop and enjoy his bounty of bed and board. From Texas came his brothers, Pitzer, James and Jeff Chisum, to help run the huge ranch; and his niece, pretty, smiling Sally Chisum, daughter of his brother James, was for years chatelaine in the big house, with a country as large as Pennsylvania over which she was the charming mistress.

Yet while he lived in rude magnificence and knew his herds grazed a long three days' ride up and down the river, John Chisum had a gnawing worry—rustlers. First it was Indians. In his early days on the Pecos, Apaches and Comanches frequently ran off stock by the thousands of head from Chisum's range. Mexican marauders swept up from below the border and drove away cattle and horses. One Mexican gang escaped with twelve hundred horses, but Chisum, pursuing with only three men, caught up with the thieves and slew three of them, whereupon the others were constrained to gallop for their lives and Chisum returned in triumph with the entire herd. Except for a time of action like this, the King of the Pecos did not carry a six-shooter.

White rustlers, however, were the greatest threat. Chisum operated on an impressive scale. In a single two-year period he marketed five thousand cattle at Tucson, six thousand at the San Carlos agency in Arizona, four thousand on the Gila River, and six thousand in Dodge City—cattle worth more than $300,000. And that was by no means all of his business during the biennium. With such large numbers of cattle being handled, the rustlers had a peerless opportunity to prey on Chisum's herds, but all the outlaw depredations did not compare with one single stroke, perfectly legal, in which for once the canny King of the Pecos was outwitted.

It was in 1876, after he had moved to the South Spring ranch, that this misfortune befell him. As stated, there had been loose accounting for some of the cattle Chisum brought from Texas, and in 1875 he faced a lawsuit over part of the herds he allegedly transported without payment. Tom Catron, a young and aggressive lawyer of Las Vegas, filed the case. Chisum was arrested and even held in jail for a short time on fraud charges, but he produced from a long zinc tube his carefully preserved powers of attorney, and the lawyer had to dismiss the suit. Catron disliked having the laugh against him. The day came when he evened the score.

In the Dodge City *Globe*, April 6, 1876, appeared an innocent news item stating that one Jesse Evans was departing from Dodge City with fifty cowboys to gather up from the Pecos Valley a herd of twenty thousand cattle which were to be purchased from John Chisum by the Kansas City cattle commission firm of Hunter & Evans. The newspaper did not add that Jesse Evans' fifty "cowboys" were fifty of the deadliest six-shooter experts who could be gathered from the turbulent Dodge City area. Nor did it recount under what terms the twenty thousand cattle were to be purchased. About this there is some dispute, but the facts as nearly as they can be learned are as follows:

Robert D. Hunter, head of the Kansas City commission house, had seen an opportunity to cash in on Chisum's old packing house and cattle debts. As cunning as the old Pecos king himself, Hunter made a quiet trip into Texas and there purchased as many of Chisum's notes and the judgments against him as possible, paying ten cents on the dollar. Then he retained Tom Catron as his attorney and rode south with Jesse Evans and his fifty fighting men.

Chisum proved quite willing to sell Hunter twenty thousand cattle at the price offered. He and his cowboys even helped gather them. After many dusty days on the Pecos flats the immense herd was rounded up and, with Evans' cowboys trail-driving it, started northeast toward the Texas border and eventually Dodge City.

With Chisum, Hunter rode to Las Vegas to settle up. The owner of the Jingle Bob may have had an inkling of what was about to happen when Hunter took him to Catron's office. It is hardly likely, however, that he anticipated the full extent of Hunter's scheme. Carefully provid-

ing witnesses to the transaction, and in the presence of the
attorney, Hunter placed an old satchel on a desk.

"There, Chisum, is your pay," he said. The satchel was
opened. It was full of the old and almost forgotten Texas
obligations of the cattleman.

Just what took place in the room following presentation
of the ancient notes, is unfortunately not of record. John
Chisum was a hard man and a brave man. He was also a
man given to pungency of expression. But he did not carry
firearms, as we have seen, except on extraordinary occasions, in which eccentricity he was joined in the West by
Branch Isbell, John Iliff, and Murdo MacKenzie, but by
precious few others. On this occasion he was unarmed and
there was, therefore, no shooting.

The legal technicalities were on Hunter's side, and Hunter was also a cool, courageous man, for all his suavity;
a man who refused to be bluffed. Even so Chisum, after his
first explosion, might have rushed out, rallied his riders,
and set out after the disappearing herd to hold it and settle
the case with a long battle in the courts. But another decisive factor intervened. Those were no ordinary cowboys
riding under Jesse Evans. They were fifty of Dodge City's
most potent killers and they carried, swinging in the holsters at their hips, certificates of their ability to hold the
cattle against any but the most overwhelming pursuing
force.

And so the long, lean King of the Pecos helplessly
watched his twenty thousand cattle driven toward the
Texas line and forever beyond his reach. Chester Evans,
now living at Lebo, Kansas, was a resident of Dodge City
at the time this occurred. He well remembers the excitement surrounding Hunter's bold stroke. The cattle were
hurried across the Texas line, he told me recently, and
driven as a single herd as far as the Indian Territory—
surely the largest single trail herd on record. There the
cows, calves, and other stockers were cut out and driven
hurriedly north to Cheyenne, to be sold, some of them

going as far north as Montana. The beef steers were shipped by rail, not to Kansas City, Hunter's home office, but directly through to Chicago.

"Hunter was anxious to get those cattle out of the country—clear out, where Chisum couldn't reach them," Evans said. "He didn't feel safe about them until there were two or three state lines between them and the Jingle Bob boss."

The story of Hunter's "legal rustle" of Chisum's cows was widely told in the Southwest and hugely enjoyed, but the dour owner of the Jingle Bob never saw much humor in it. In fact, he later sought to deny it, saying he sold willingly to Hunter "because he wanted to get out of the business." If so, he changed his mind, for he remained in the business until his death in 1884.

Even a coup as gigantic as that of Hunter did not stop John Chisum. His herds continued to increase and again he turned his attention to his immediate problems. Chief of these was the constant theft of his stock by rustlers who made no pretense of legality, and who were much nearer the Jingle Bob range than the astute Kansas City speculator.

About the time Chisum settled at the Bosque Grande, Major L. G. Murphy established himself in the mountains near the Mescalero Apache reservation, at the little town of Lincoln, New Mexico. Murphy came to the Southwest with General J. H. Carleton's California volunteers during the War between the States. For a time he ran a sutler's store at Fort Stanton, but sold out, reportedly because the military authorities disapproved of some of his business methods. By 1875, however, Major Murphy was prospering mightily at Lincoln. He owned the leading store there and operated, besides, a flour mill, a hotel, a saloon—and a cattle ranch. The cattle ranch was a mystery which puzzled every cattleman in New Mexico. Murphy had it first on the Carrizoza plains, but later moved it to the Seven Rivers country—close, uncomfortably close, to the Jingle

Bob holdings. And it was observed by cowmen that although Murphy's cattle never numbered more than three thousand at a time, he somehow managed each year to sell thousands of steers. Either his cows were endowed with amazing fecundity, or else . . .

To John Chisum the alternative was clearly apparent. He charged Murphy, without mincing sentences, with stealing his cattle.

In the West there were three fighting words. Call a man any of them and reach for your gun, because it was a certainty that he would be going for his. Those words were liar, coward, and thief. Chisum applied them all to Murphy, outspokenly, to anyone who cared to listen. The ex-sutler only laughed. What did a little insulting language mean to him? Greatly was he profiting and he could afford to ignore the Chisum mouthings. Besides he was building up power in another direction through which some day he might make Chisum very sorry for those statements.

Murphy was becoming a political power. His town of Lincoln was the county seat of Lincoln County, which then contained four present-day New Mexico counties— Lincoln, Chavez, Eddy, and Otura—and part of a fifth— Doña Ana. It included one-fifth of New Mexico, a quite populous mountain country cut off from outside civilization by barriers of desert. Most of this isolated territory went to Lincoln to trade, and most of the trade went to Murphy's store. As the big man of Lincoln the major controlled lesser men. His dollars were influential in politics as dollars have a way of being. The sheriff of Lincoln County, for example—James A. Brady—was known to be a Murphy tool. A young lawyer, Alexander A. McSween, fresh from Atchison, Kansas, with his bride, was retained by Murphy as his attorney in anticipation of possible legal complications.

Month after month, winter and summer, the rustlers wrought havoc in Chisum's herds. Faithful and courageous though the Jingle Bob cowboys were, they could not watch at once all the frontiers of that valley kingdom. So Chisum, on what he considered good evidence, swore out warrants for men he suspected of stealing his cattle. Sometimes Sheriff Brady went through the motions of arresting the men named in the warrants, but it was amusing to observe how incapable was the jail at Lincoln of holding them. Invariably the prisoners shortly escaped, either through their own efforts, or through the aid of friends, who rode into town, made a noisy show of six-shooters and Winchesters, and then received from the complaisant jailer of the Murphy faction the keys of the prison.

There came a day, however, when Chisum obtained evidence impossible to controvert, and on it swore out warrants. The accused rustlers were arrested and bound over for trial. The story is that Murphy ordered McSween, the young Kansas lawyer, to defend the rustlers, and that McSween, a man of devout religious convictions, refused on the ground that the men were guilty and it was against his conscience to do so. It is known that McSween was a Presbyterian who had started to prepare for the ministry before he took up the study of law; also that shortly after this episode the lawyer became Chisum's legal representative in Lincoln.

To Murphy's thinking this was a thundering outrage, and the more so when he learned that McSween was entering the mercantile business in Lincoln as a competitor to Murphy's store. Hitherto the lucrative trade had been a virtual monopoly of the major's. He fixed the prices on whiskey, air-tights (the Western name for canned goods), saddles, ammunition, soguns, boots, lariats, six-shooters, Winchesters, hats, tobacco, and all the other necessities of the cow country, because he practically controlled the supply.

McSween was financed in his business undertaking by a young Englishman, J. H. Tunstall, one of that wave of

Britons who, enamored of the American West, came to this country to live in the romantic wilderness of cattle. Addicted to riding breeches and sporting caps, and full of "By Joves" and "My words," Tunstall greatly diverted the cowboys, but he went along apparently oblivious of the laughter behind his back, bought a ranch on the Rio Féliz, thirty miles from Lincoln, and by his whole-hearted friendliness won the respect of most men and the love of one. The McSween-Tunstall store began to make its bid for trade in Lincoln. And now the hand of Chisum showed itself. A bank opened in the rear of McSween's store building. Chisum's name appeared as its president. It has been claimed that the King of the Pecos actually had no money invested. No matter; the mere fact that his name appeared on the bank stationery was tantamount to a declaration of war. He was carrying the fight to his enemy's country and this further infuriated Murphy.

Still further intensification of the feud came after the death of Colonel Emil Fritz, a former partner of Murphy's, who left an estate that included ten thousand dollars in insurance. To this money Murphy laid claim on the ground that Fritz owed him a debt. McSween, representing the Fritz heirs, rejected the claims and was counter-charged by Murphy with embezzling from the estate. He was even placed in jail in Las Vegas in January, 1878, but was released on bail in time to lose his life and win an unforgettable name for futility in the leadership of fighting men.

To collect on the Fritz estate, Murphy meantime sought to assess against McSween's property. He was informed that McSween had turned his assets over to his partner Tunstall to avoid an attachment. Thus far the lawyer had played the game according to the legal code he knew, and had done rather well. But he was far from the civilization in which his kind functioned best. The time was almost upon him when he should see all his knowledge of jurisprudence and his belief in orderly processes tossed into the rubbish heap by raw, red direct action and elemental savagery.

In the new phase of the struggle, the first blow was struck by Murphy. He sent a posse, headed by Billy Morton, one of Sheriff Brady's toughest deputies, to the Rio Féliz to take over Tunstall's property in lieu of the money Murphy said was owed him. A hard-riding cowboy brought word to Dick Brewer, foreman of the Rio Féliz ranch, that Morton and more than twenty armed men were riding his way. Brewer smelled death in that visitation. He urged Tunstall to leave, but sturdy in his British sense of constitutional legality, the Englishman refused. Hours of argument ensued and finally, unwillingly, Tunstall agreed to go. He was accompanied from the ranch by Brewer, a deputy marshal named Robert A. Widermann, and a slight, unknown boy named William Bonney, who was to achieve fame as Billy the Kid.

Into the Rio Féliz ranch swept the Morton posse and began gathering Tunstall's cattle to satisfy Murphy's claim. Part of the men, however, took up the trail of the Englishman himself. They had been drinking and were irresponsible. Tunstall disliked running away. He hung back to speak to his adversaries, against the horrified protests of Brewer, Billy the Kid and Widermann, who were devoid of any illusions regarding the posse. Seeing it was impossible to convince the Englishman, the three others finally rode on ahead, while the Briton stopped to converse with the pursuing mob. Very shortly Tunstall paid the penalty of his fatuousness. Shots sputtered out. Beside his dead horse the bluff Englishman, who had never harmed or even spoken an unfriendly word with any of these men, pitched to the ground, dead. The Morton posse, which included Frank Baker, Billy Matthews, Tom Hill, and John Robinson, among others, rode away.

On a distant hill a boy sat his horse and watched with a wild glare the men leaving after the atrocious murder. Billy the Kid had idolized Tunstall. They were opposites in the social scale—the ignorant, undersized border waif, and the cultured, handsome and wealthy English gentleman. But Tunstall, kindly and generous, had befriended the Kid, and the Kid loved the Briton as he never loved any other being before or afterward.

XXXIII

DESTROYING ANGEL WITH SPURS

There was nothing prepossessing or impressive about the appearance of Billy the Kid—certainly nothing to warn one that here was, in many respects, the most notable six-gun killer of them all. A man who knew him well, Frank Collinson of El Paso, once described him thus:

"A slight, boyish-looking chap, badly weather-beaten . . . not much to look at. About five feet seven or eight inches, not over one hundred and forty pounds at the outside. Rather sloping shoulders, no chin, good nose, very good blue-gray eyes . . . everything he wore, from his old black hat to his boots couldn't have cost ten dollars all together when new."

The Kid's history has been traced by Walter Noble Burns and others. He was born in New York's Bowery, christened William Bonney, and was taken at an early age by his parents to Coffeyville, Kansas, a tough border town where later the Dalton bandit gang was wiped out in a street fight. After his father's death, his mother took him to the mountains where, in Colorado, she married a man named Antrim—from which circumstance Billy, in his early years, was often called Billy Antrim. He is said to have killed his first man at twelve—a blacksmith in Silver City, New Mexico, who spoke slightingly of the boy's mother. After the killing, Billy bade his mother good-by and fled from the country. Forever afterward he was an Ishmaelite; mother and son never met again.

Even at that early age he was an expert gambler, a prodigy with the cards as he later became with the six-shooter. If he had any definite occupation in his brief life, aside from gunman and cattle rustler, it was as a gambler. In the next several years he roamed the Southwest, gambling, dancing with the *señoritas*, killing a man here or there in quarrels, hiding out in the mountains, then reappearing suddenly in dance hall or gaming den—a deadly youth whose moods and technique with the gun men learned to dread and respect.

And with reason. In the lurid gallery of super-killers in the West, Billy the Kid stands out in sharp relief, eclipsing all, with the possible exception of Wild Bill Hickok. Occasionally, as the law or circumstances required, Hickok took his toll and cut his notches, omitting Indians. His list was longer than the Kid's, in fact two or three others had longer lists than Billy's, but those lists were compiled over much longer periods. Considering the Kid's age and the brief era of his fighting days, Wild Bill's fame as an executive of the six-shooter pales before that of this young genius of annihilation. With Billy the dealing of death was a high art, and necessarily so, for after his first half-dozen preliminary assassinations it became a case of kill or be killed.

Nor are the hues of romance wanting in him. He had his good points and his friends—some of them powerful ones. Women often idolized him and many men remained faithful to him to the death. He possessed a certain oblique code of honor, made good his word whether it was a threat or a promise, enjoyed the life he lived, and was withal as genial and smiling a youth as ever blasted out a man's life in cold blood. Still living in the Pecos country are gray-haired men and women who revere his name. In a superlative degree he had the virtues that counted in the wild country of cattle—nerve and courage.

Wherefore the legend of Billy the Kid is still today as fresh and vivid in the Southwest as when he was riding there. In Spanish and English his exploits are celebrated over and over, and *El Chivato*, as the Mexicans call him, has become a symbol of crude but nonetheless gallant frontier knight-errantry.

This was the youth who stood above Tunstall's grave, after McSween caused the body to be brought in and buried. Standing there he said: "I will kill every one of the men that had a hand in this murder, or die trying. You were the only man that ever treated me like I was freeborn and white."

It was the wail of a child, bewildered and hurt. It was also the snarl of a savage animal.

In Lincoln County war began. Unwilling from religious scruples to participate in violence and inadequate for leadership in a lawless society, the pathetic McSween found himself pushed into a prominence all undesired. Constantly and ineffectively the lawyer, who had once dreamed of being a minister, pleaded for peace. To the last he refused to carry a gun—a circumstance which did not prevent him from dying by one.

As Chisum's representative in Lincoln, McSween induced John P. Wilson, justice of the peace, to appoint Dick Brewer, Tunstall's foreman, constable. Sheriff Brady and his whole force of deputies belonged heart and soul to Murphy. It was necessary to have an official arm to counter him. But had McSween known how his legalistic move would escape him and become an engine of bloody destruction, he probably never would have asked for the appointment of a constable. Brewer accepted the commission eagerly. He was intent on revenge for Tunstall and he at once deputized a collection of the choicest desperadoes in the country to hunt down the Englishman's murderers. In this group rode the Kid, Charlie Bowdre, Tom O'Folliard, Doc Skurlock, Hendry Brown, Jim French, John Middleton, Fred Wayte, Sam Smith, Frank McNab and George and Frank Coe.

This posse struck almost at once. After a running fight in the Rio Penasco valley it captured Billy Morton and Frank Baker—the leader and a member of the Tunstall murder posse. Only after Brewer promised them safety did the men surrender, but neither placed any faith in the promise. That night they were guarded at Chisum's South Spring ranch and next morning were told to mount for the ride to Lincoln. As they did so one of them gloomily predicted to his comrade that neither would reach his destination alive. One of the possemen, McClosky, overheard them and reassured them:

"They can't kill you boys while I'm alive."

It was a speech either courageous or foolhardy since McClosky must have known the feeling among the men with him. It meant his own death. Riding down the Bonito canyon toward Roswell, as they were passing through a gulch near Agua Negro, one of the possemen rode up to McClosky and coldly shot him dead from his horse. At the sound of the gun Billy the Kid turned in his saddle in time to see the prisoners, Morton and Baker, spin their ponies about and spur frantically for freedom. He whipped out a revolver and fired, rather negligently, twice. With three riderless horses the posse continued on its way. Dead in the canyon lay Morton, Baker, and McClosky, the deputy who sought to protect them.

The Kid felt justified in his two killings. Definitely the men were known to have been chiefly responsible for the killing of Tunstall. Besides they were trying to escape. It was the middle of February, 1878, when the Bonito canyon shootings occurred. On the morning of April 1 came the next blood letting. As Sheriff Brady, Deputy George Hindman, Billy Matthews the court clerk, and George

Peppin—all Murphy men—walked past the McSween store on their way to open court, a volley of shots suddenly clattered against the narrow canyon walls that towered on both sides above the little town.

In an instant it was done. Sheriff Brady, who sent out the posse that murdered Tunstall, was dead in the dust. Hindman, mortally wounded, crawled in a trail of blood toward the door of the little church of San Juan. Matthews and Peppin scuttled for safety around a corner. The shots had been fired from behind the adobe wall that surrounded the patio of the McSween home.

While the smoke still drifted in a gray veil down the street, the heads of Billy the Kid and some companions appeared over the wall, necks craning to see what execution they had done. Billy's eyes, as they rested on the body of Brady, his enemy, lying in the dusty road, lit up. Over the wall the young killer clambered. Sheriff Brady was carrying a fine rifle and revolver when he died and these lay beside his body. The Kid picked them up and with never another glance at the dead man, took them back with him over the wall.

Major Murphy was enraged and frightened by the two sets of killings. He set in motion every force he could control to catch the murderers. McSween, W. P. Shields, and Widermann were arrested but they were not even in town when Brady and Hindman were killed, and were speedily released. Meantime the anti-Murphy forces, temporarily in the ascendant, had made John N. Copeland sheriff, but Murphy, using his influence with Governor Samuel B. Axtell, quickly overturned this and secured the appointment of George Peppin instead.

Still, however, Dick Brewer rode as constable, and on April 14 he led thirteen men to Blazer's mill, in the Mescalero reservation, to arrest Andrew J. Roberts, locally known as "Buckshot," because of the amount of lead he carried in his body due to a varied career as a Texas ranger and a United States soldier. Just why Brewer wished to eliminate Roberts is not clear; it has been said the real object of the expedition was someone else but Roberts came in the way. The little ex-ranger, partly crippled but game, did not appear to understand it either, but he did understand that here were men inimical to him and he fought like an old, scarred wolf for his life. He and Charlie Bowdre fired at the same instant, Bowdre's bullet passing through Roberts' body, while Roberts' bullet tore off the buckle of Bowdre's cartridge belt. With a death wound the ex-ranger staggered back and shot Jack Middleton just over the heart. Another shot and George Coe had lost his trigger finger. He carries that memento of Buckshot Roberts to the present day.

The posse dove for cover. Roberts was shot through the bowels and bleeding to death, but he dragged a mattress to a window and settled down to fight until he died. Behind a pile of sawlogs a hundred yards from the mill, lay Dick Brewer, the constable. He raised his head to shoot and Roberts' buffalo gun roared; the huge slug took off the top of the posse leader's head. After that the attackers retreated, carrying their dead and wounded. Undisputed master of the field, Buckshot Roberts slowly sank and died.

Expert gunmen—professional killers—had an economic place in the frontier West. They turned up wherever there was trouble, and trouble was pretty constantly fermenting in one part or another of the range country, sometimes in several places at a time. Like all mercenaries, they espoused the side which made them the first or best offer, and thereafter made that cause their own, even if it meant the murder of a close friend who shortly before had been a trusted scouting companion but now was aligned with the other side for his bread and butter. In every well-developed cow country difficulty the best of the fighters came to the top. So it was in Lincoln County after the Roberts fight. Dick Brewer, chief of the McSween feudists, was dead. As if by divine right Billy the Kid stepped into the leadership left vacant. He was the youngest of them all, eighteen years old and small for his age; yet already his name was a synonym for calculating deadliness and audacity in assassination unparalleled on the lawless frontier.

The Murphy forces, meantime, were being strengthened by outside fighters and in the following July the inevitable showdown came. Frank McNab, a Chisum foreman, with Jim Saunders and Frank Coe, were riding down the Bonito canyon, when Sheriff Peppin and a posse jumped them. Up and down the gorge the bullets began skipping in lively fashion. Presently McNab, fatally wounded, crawled off into the brush to die. A bullet broke Saunders' ankle and he and Coe were captured. With its prisoners the posse rode to Lincoln but a heavy volley greeted it from the town—a big force of McSween men already occupied Lincoln. At a quarter of a mile, George Coe, brother of Frank, shooting without his trigger finger, broke a leg for Bill Campbell of the Peppin forces. In the excitement Frank Coe escaped and joined his friends while Peppin and his posse withdrew into the hills.

Billy the Kid was not in Lincoln during that fight. He was at John Chisum's South Springs ranch in the Pecos valley. There he and five or six followers were surrounded by a force of men under Marion Turner, one of Peppin's deputies. For a time the guns crackled, but it was a long-range, bloodless battle, which ended when a daring Jingle Bob cowboy rode the gauntlet of bullets and brought to the rescue twenty-five of Chisum's ranch hands from a roundup camp a few miles away. Turner's posse departed, the dust of their ponies' hoofs indicating their haste, and the Kid and his men rode to join their friends at Lincoln.

The following night the Peppin forces, sixty strong, silently surrounded the sleeping town. In the McSween house lay the Kid and a handful of his men; McSween himself and most of his followers were absent. Early in the morning Billy was awakened by a hail from the sur-

rounding hills. It was a call from Peppin informing him
that he and his men were surrounded and ordering them
to surrender. The Kid laughed like a wolf at the order.
Before the sheriff could act shrill whoops sounded down
the Bonito canyon, followed by a clatter of horses' hoofs,
and McSween with thirty-five men rode in.

To the very last the lawyer hoped for a peaceful adjudication, but he was dealing now with forces beyond his
remotest conception. The first shots sputtered out even
as he counseled moderation and temporizing; it was the
opening of a battle which has become as famous as any
in Western history.

No casualties at first; both sides were too adept at using
cover. A tame fight, the Kid thought, and he yawned as
he waited for "a good square crack at anybody," as he
put it. McSween, utterly unnerved, still implored everyone who would listen to him to trust in the Lord and
maintain the peace.

"Go ahead and trust in the Lord," Billy told him, half-
amused, half sardonic. "The rest of us will trust our six-
shooters."

Night came at length to end the shooting which had
been annoying but so far futile on both sides. In the dark
hours two of Peppin's sharpshooters, Lucio Montoya and
Charlie Crawford, made plans to do better execution on
the morrow. They climbed up behind the small peak you
can see today lifting its head close above the town. From
the rocks at its top they could gaze down into the patio
of the McSween home, and as the sun poked its rim above
the mountains at dawn, a man crossed the compound.
Both sharpshooters fired at him, and both missed.

Down in the patio lay Fernando Herrera, an old Mexican buffalo hunter with a reputation for great skill with
his heavy Sharp's rifle, who had espoused the cause of "the
Keed." Waiting until Crawford raised his head above the
rock which loomed overhead and a quarter of a mile away,
Herrera pulled his trigger. It was a miracle shot. The

bullet drilled Crawford through the skull and the Murphy partisan shrieked once, then rolled all the way down the mountain, dead. At his companion's yell, Montoya forgot caution and leaned over his concealing boulder to see what had happened. A second bullet from Herrera shattered his leg.

The narrow, crooked street which was Lincoln's sole thoroughfare was now curtained with gray powder smoke and the hills all about echoed deafeningly to the explosions of a hundred guns. Herrera's shots were the preliminary to a general action in which every man on each side poured bullets as fast as he could at the enemy. Yet after those two first casualties no more were wounded that day, so close to cover did the fighters of both little armies lie.

Once more night intervened and again on the third day firing was resumed. But it ceased late in the day; a faint bugle call had been heard. While the contending feudists waited in surprise, two companies of Negro cavalry, accompanied by two Gatling guns, came jingling down Lincoln's street. At their head was Colonel N. A. M. Dudley of Fort Stanton, to whom had been carried word of the battle between the cowmen.

Dudley conferred with the respective leaders. His conclusion was that the *de facto* sheriff—Peppin—was in charge of a perfectly legal posse, which was attacking a band of men at least technically outside the law. Under the circumstances the officer could do nothing. He warned the leaders of both factions, but he was powerless to interfere unless called upon formally by the proper authorities. In this case the "proper authorities" seemed in control, so Dudley withdrew his men and camped.

One purpose had been served by the parley—it gave Peppin's forces a great advantage because, under cover of the truce, some of the besiegers crept to the rear of McSween's house and succeeded in setting it on fire. Now as the battle reopened, the flames began to rise and the surprised defenders fought them vainly. During this period reinforcements tried to enter the town to join the McSween forces, but Dudley intervened with his cavalry and prevented it.

As darkness descended it was apparent the defense was about finished. Either the beleaguered men must surrender, escape, or be roasted alive. Surrender was out of the question—everyone knew what would be the fate of prisoners. Escape was the only alternative. All about lay the rifles of the enemy. Mrs. McSween and two other women in the house were sent forth first. They walked boldly out and not a shot was fired. The West did not make war on women.

It was a different situation, however, that the men faced. Whenever one of them showed his head, a shot instantly rang out and a bullet whistled perilously near. The house, however, was becoming untenable.

Harvey Morris and Francisco Semora made the first dash from the flaming building. In the glare of the fire they were sharply silhouetted and both were shot dead before they had taken more than a few steps. Vincente Romero was the next killed. Now McSween, still clinging to his belief that if he did not carry a gun he would not meet violence, stepped forward, carrying his Bible like a talisman and calling, "Gentlemen, I am McSween." That was all the Murphy men wanted to know. A dozen rifles flashed in the darkness and the Kansas lawyer crumpled over, dead.

Then the defenders came scuttling forth like rats from a burning barn—so rapidly and in such numbers that they seemed to confuse the riflemen waiting for them. Two of them were wounded, but the others, ducking, zigzagging, dodging, escaped the whining bullets and tumbled over the wall into the safety of the brush of the river below. Last of all was Billy the Kid. Like a captain on a burning ship, he saw that every man was gone before he left. The rafters were almost ready to collapse and the roof was a burning

inferno, when suddenly he appeared in the flaming door-
way, with a six-shooter in each hand. Across the thirty-
foot patio he sprinted, bullets thudding about him, his own
guns blazing. Bob Beckwith, the man who fired the shot
that was fatal to McSween, was himself instantly killed
by the Kid's deadly marksmanship. Two other Murphy
men fell back, wounded. Then Billy was gone. Sheer nerve
and his murderous skill with his guns had enabled him to
escape over the wall and down the bank into the underbrush of the Bonito.

The three-day fight ended the Lincoln County war and
established the ascendancy of Sheriff Peppin, while making
an outlaw of Billy the Kid. But it did not by any means
end the reign of terror in the country. Murphy was dead;
he died in Santa Fé shortly before the battle, ruined, his
ranch, store, saloon, and money all gone to pay his army
of hired mercenaries. Chisum, having seen the demise of
his enemy, lost whatever interest he originally had in the
fighting. He outlived the vendetta by only a few years,
dying in 1884 at Eureka Springs, Arkansas.

But the feud long refused to die. It grew worse, in
fact, the older it became. Cowboys were "dry gulched"
in lonely canyons and on the flats of the great wild range.
Cattle continued to be stolen and their herders died defending them—or rustlers were killed as they attempted
to run off herds. Deadly as a sidewinder rattlesnake, Billy
the Kid rode through it all, a small, slope-shouldered figure, instinct with destruction. News of the great range
feud reached at last clear to Washington, and President
Rutherford B. Hayes sent General Lew Wallace, author
of *Ben Hur*—on which he was then writing—to New
Mexico as governor with instructions to end the fighting.
Wallace arrived in the territory, sent for the Kid, and
when the latter came at his summons, offered him amnesty
if he would leave the country.

"No," said Billy shortly. "This is my country and I'm staying here."

He rode away, his revolvers impertinently glittering, and Wallace looked after him with a sigh, saying, "I believe he has the making of a man in him . . . if only he would listen to me."

But the courtly soldier-novelist failed to comprehend the code of the West and least of all that of a feudist like the Kid. He did not know the outlaw still had names on his list with whom he felt impelled to square accounts for his dead idol, Tunstall the Englishman. The greatest man-hunt ever held in the Southwest began. A six-foot, four-inch ex-buffalo hunter named Pat Garrett, who formerly had been the Kid's close friend, became sheriff and took the office as a solemn obligation to bring his former comrade to justice. With an army of picked deputies he rode the hills of New Mexico, day and night.

But for months he failed to corner the Kid, and the Kid killed and killed until his death total was nineteen. Then at last the boyish outlaw was run to earth—surrounded in an abandoned stone house by the posse of the grim, lank buffalo hunter, Garrett. He fought, but in the battle his closest friends, Charlie Bowdre and Tom O'Folliard were killed, and in the end Billy himself surrendered. He was tried at Mesilla and sentenced to be hanged at Lincoln.

Now began a lurid sequence of events. Knowing his old friend, Pat Garrett took no chances with him. While he was awaiting the execution date the sheriff confined his prisoner in a room above Murphy's old store—no chances were to be taken on that leaky jail—and the Kid sat in handcuffs and leg-irons. Also he was guarded day and night by two deputies, one of them being Bob Ollinger, an old Murphy partisan and the Kid's bitterest living enemy.

"Never let Billy see the color of your backs," was Garrett's solemn warning to his deputies. Yet out of this desperate situation the little outlaw made his escape.

Garrett one day rode to another town to make arrangements for the rope and scaffold that were to hang the prisoner. As the long hours dragged, the Kid asked J. W. Bell, one of the deputies guarding him, to play a game of cards. Ollinger yawned and presently went across the street to eat lunch. It was an old trick the Kid played. Dropping one of his cards on the floor, he stooped as if to pick it up, then quick as a striking adder dove for the guard's pistol holster. Out came the six-shooter in the outlaw's manacled hands. Bell, knowing his fate was upon him, leaped in frantic fear for a window, but fell clattering full length on the floor as the Kid shot him in the back. At the sound of the revolver, Ollinger came running across the street. He heard his name called from an upper window and looked up, his face paling with horror as he saw the Kid's vulpine grin for the last time. Almost cut in two by a blast from a double-barreled shotgun loaded with buckshot—his own weapon—Bob Ollinger rolled lifeless in the red-spattered dust. Then Billy, obtaining a Winchester from the arms rack inside, ordered the citizens of Lincoln to remove his leg-irons and handcuffs and bring him a horse. Not a move was made to stop him as he rode away.

Back into town Garrett hotly galloped, and once more the man-hunt was on. It was to the death, this time. Through the mountains the great, gaunt sheriff and his posses harried the undersized killer, back and forth. But they might never have caught up with him had it not been for the streak of romance in Billy's nature. Repeatedly his friends urged him to escape to Mexico from a territory that was becoming too hot even for him, but the Kid instead elected to pay a visit to a woman who fascinated him. She lived at Fort Sumner and it was there, through chance and the treachery of a derelict friend of the outlaw's, that Garrett found him. Accompanied by two deputies the sheriff went to the house of Pete Maxwell, where the outlaw was reported.

Greatly daring, Garrett left his men to watch outside and entered the house alone. Into Maxwell's bedroom he crept and lurked there in the darkness. From the porch came the voice of the Kid. He had stepped out in his stocking feet and noticed in the gloom the shadowy figures of two men—the waiting deputies.

"*Quién es?*" came his voice. When they did not reply, he stepped into Maxwell's room to inquire who they were. Out of the darkness blazed a gun—Garrett's six-shooter—and Billy the Kid curled up lifeless on the floor. He was twenty-one years old, and he had killed twenty-one men.

Two grayed veterans, each bearing the scars of the Lincoln County feud, keep alive the memory of the vendetta today near the scene where it was fought. They are Bob Brady, son of Sheriff James A. Brady, who was killed by the Kid at Lincoln, and George W. Coe, one of the Kid's loyal followers. Brady wears a mustache to hide the disfiguring scar where one of Billy's .45 slugs ripped his mouth. Coe is minus the trigger finger on his right hand —carried away by Buckshot Roberts' bullet in the fight at Blazer's mill.

Back in the '70's these men expressed their differences with smoking revolvers. Today, with the years sitting heavily on both pairs of shoulders, the appeal to bullets is forgotten and except when the subject of the Kid comes up the old fighters get along well together. Bob Brady lives in a neat adobe cabin at Hondo, near the fork of the Bonito and Hondo Rivers. Seven miles up the Hondo valley is a roadside store, operated by George Coe, where he loves to lie in his hammock and talk to visitors. Ten miles up the other fork, in the Bonito valley, lies Lincoln, no longer a lusty, roaring cow town, but a decaying little hamlet, peopled more by ghosts than living persons.

There you can still see the round stone tower called

El Torreon, where the settlers fought off the Apache Indians. McSween's store, its shutters lined with steel, still stands. One end of it is now used as a store, the remainder as a warehouse. A graveyard up the hill contains the headmarkers of more than a few men who died with their boots on. In its place also is the little church of San Juan where George Hindman dragged his bleeding body up the steps and died after he and Sheriff Brady were ambushed in front of McSween's. The open space where five men met death the night McSween's home went up in flames is now the back yard of the home of Julio Sales. And up the street is Juan Patron's store, where Billy used to loaf. On the lintel of its door are three letters, carved crudely by an illiterate boyish hand: K I D.

XXXIV

WYATT EARP PLAYS OUT HIS HAND

Arizona, in 1880, was the abode of the toughest characters in the West—and that meant in the world. This was due to a combination of circumstances. Other Western states were indulging, almost simultaneously, in a purge looking toward a new and hitherto unsought state of purity and righteousness. One after another the lawless cow towns of Kansas became respectable until only Dodge City remained—and the "better element" was taking charge of it. In Missouri the Jesse James gang was breaking up and the leaders being hunted to their death so that to many vivid individuals the state was becoming vastly insalubrious. The rangers in Texas were making conditions most uncomfortable for the unlawful, even in the wild reaches of the Staked Plains and the Big Bend. A ranger was a bad man to have on your trail, so there was an exodus of undesirables from Texas. Finally, in New Mexico, as an aftermath of the Lincoln County war, Pat Garrett was sweeping the country clear of both sides in the Billy the Kid feud, and a considerable portion of the population was finding it convenient and even necessary to go hurriedly elsewhere.

Meantime, in Southern Arizona, Ed Schieffelin had found the Tombstone diggings and the famous mining capital of that name sprang up. Almost immediately came a vast mushroom growth of mining activity at Tombstone, Bisbee, and elsewhere in that section of sun-blasted mountains and flats. This meant a market for beef. In Cochise County, which is the southeastern corner of Arizona, where wide valleys run between the Peloncillo and the Chiricahua Mountains on the one side and between the

Chiricahuas and the Dragoons on the other, cattle activity therefore began.

At first the ranches were operated by legitimate cattle-men from Texas, who trailed herds across from the Pecos and established themselves in the canyons and along the stream-beds, wherever water could be found. Later, how-ever, the district became headquarters for a numerous and malodorous collection of outlaws, murderers, and thieves. It was an ideal hiding place for men who for various rea-sons wished to be lost to the world, and to these men cattle rustling was a natural and congenial occupation.

For two reasons the honest ranchers did not bother them at first: The rustlers were extremely dangerous to inter-fere with; and in the beginning they directed their activi-ties chiefly at the Mexican ranches south of the border. This last was easy and simple. Cochise County lay on the international boundary. It formed a sort of *terra incognita*, where men who usually answered to names not given them at birth lurked in hidden arroyos and watched every stranger who entered. In that country a man took good care that nobody rode behind him and kept warily away from those he met, breathing deep with relief when he was out of the baleful area's mysterious menace. From this lost land it was easy for hard-riding rustlers to go far south, round up cattle in Mexico, and bring them back to some hidden corral where they could be rebranded before being sold in one of the mining towns.

Even to honest Texans this practice did not seem to be particularly iniquitous. Very deep lay the hatred of a Texan for a "greaser"—based on the cruelties of the Texas revolution and the events succeeding it, the early activities of Mexican rustlers, and their noxious alliance with the raiding Comanches. What happened to a Mexican, there-fore, was a matter of immense indifference to a Texan—so long as it was unpleasant—and the ranchers merely grunted and rolled a cigarette when told of depredations south of the border.

Two ranches were the chief headquarters of the cow thieves of Cochise County—the Clanton ranch in the San Pedro valley, and the McLowery ranch in the Sulphur Springs valley. Countless smaller establishments were scattered in the barren hills. The Clanton clan was as rough and tough as ever the West bred. It was led by its patriarch, who bore the initials N.H., but was never spoken of by friend or foe otherwise than as Old Man Clanton—who had been run out of California in the gold rush days by the vigilantes. Passing well he knew the use of the business end of a six-shooter and also of a running iron to alter other men's brands. With their harsh old sire rode three hard-eyed, hard-riding, hard-shooting sons: Ike, Phin and Billy, all born and reared in outlawry. The McLowery ranch was operated by brothers, Frank and Tom, both dangerous men with weapons. Under the leadership of the Clantons and McLowerys operated a malign group of desperadoes. Curly Bill Brocius, John Ringo, Joe Hill, Jim Hughes, Pony Deal, Frank Stilwell, Jim Crane, Zwing Hunt, Harry Head, Billy Leonard, and Billy Grounds were some of them. All told, probably three hundred cutthroats lived in this rustlers' paradise, all reckless, all killers, all utterly without compunctions.

For a time, as was said, the legitimate ranchers minded their own business, and it did not violate their convictions on occasion to purchase a few "wet" steers or horses. "Wet" was a border term for animals that had been swum across the Rio Grande into Texas, and hence had come to apply to all stolen stock brought north across the Mexican border, although in Arizona there was little dampness to justify the name. After a while, however, raiding grew less easy and profitable in Mexico. The Mexican ranchers fought back. On the trails of rustling bands swarmed little waspish *vaqueros*, and men died in the cactus wastes. Finally a major disaster occurred. Curly Bill Brocius and several rustlers had stolen a herd of Mexican steers and were bringing them north to the Clanton ranch, when they were overtaken. The rustlers killed a dozen or so Mexicans in a negligent way, then brought along the cattle. But the *vaqueros* were not through; they did not intend to let the stolen herd ever reach market. After working the cattle through his branding corrals, Old Man Clanton, with five men, started to drive the stolen stock to Tombstone. In the passes of the Guadalupe Mountains the dogged Mexicans ambushed them and killed Clanton and four of his men. Only one, Harry Earnshaw, escaped. The recovered cattle were driven south by the victorious *vaqueros*.

Such incidents reduced the popularity of raids into Mexico and the rustlers thereafter turned their attention to cattle companies in Arizona. In this they were joined by the Mexicans who, having repelled, so to speak, a *gringo* invasion, now counter-attacked with zeal. Nor did the outlaws of Cochise County limit themselves to cattle theft. Men were robbed and murdered. Stages were held up. Law officers were intimidated or corrupted. The whole section of Arizona was gripped by a reign of terror.

But the criminals' rule broke unexpectedly. It broke on a stubborn rock—the loyalty to one another of four brothers, and of a friend of the four.

Wyatt, Morgan, Virgil and Warren Earp were remarkable even on the border which was full of remarkable men. All of them were fighters, tall, mustached, with twisted-wire wrinkles about their eyes from long riding in the sun, and the natural, bow-legged cowboy walk. They were fond of one another and intensely loyal, so much so that they preferred the company of their brothers to that of other men. Wyatt was the leader, a famous gun-fighter already, with a background of service as a peace officer at Ellsworth, Wichita, Dodge City and elsewhere. In the group was one outsider who somehow reached an intimate footing with the brothers. This was John H. Holliday, a dentist, known always as Doc, a small, consumptive, coughing man, deadly as a cobra.

Curly Bill Brocius stepped into the leadership of the rustlers after Old Man Clanton's death. One of the collateral
activities of the gang was "standing up" the stages which
often carried gold from Tombstone and Bisbee. So strong
was the outlaw element that it influenced, if it did not actually control, the law-enforcement bodies in the country,
with the exception of the Earps. Wyatt Earp had ridden
shotgun with the stage line and had served as deputy under
Sheriff Johnny Behan, until he became disgusted and quit
to take a job as a deputy United States marshal. He had,
moreover, won the enmity of the rustler crowd when he
took from Billy Clanton, at the point of a gun, a stolen
horse.

Concerning just what took place thereafter, there is a
wide divergence in stories in the West. Billy Breakenridge,
who became sheriff of Cochise County and was deputy at
the time of the Earp-Clanton feud, wrote a book entitled
Helldorado, in which he referred to the Brocius-Clanton-
McLowery crowd merely as "cowboys" and charged that
the Earps, notably Wyatt and Doc Holliday, were aggres-
sors, mistreating the inoffensive sons of the saddle until
they were goaded into fatal action. The opposite viewpoint is taken by Stuart Lake in his *Wyatt Earp, Frontier
Marshal:* that Earp, standing for law and order, incurred
the enmity of the rustling contingent by arresting some of
its members, and fought to the finish with the outlaws on
that issue. Even today in the Southwest you will find those
same differences and all the shades of feeling between
them. In the ghost town that Tombstone now is, I put the
question a year or so ago to a store keeper and an eating
house operator, two of the few business people left there.
From one I received the answer: "Wyatt Earp was the
best and bravest officer who ever wore a star in the West."
From the other: "Wyatt Earp was a yellow-livered sneak
and murderer. He killed for his own interests." These
opinions, held by persons living in Tombstone, operating
businesses near each other, and delivered only a few min-

utes apart, illustrate the complete lack of unanimity of belief existing in Southern Arizona, which is today nearly as heated over the question as in the heyday of Tombstone.

Yet an examination of events both before and after the feud must convince the fair-minded that the Clanton-McLowery element was, as the Earps contended, made up of enemies to law and order. Nearly every prominent member of it was killed or sent to prison for criminal activities. And it is certain that the Earp campaign gave a terrific set-back to outlawry in Cochise County.

On a hot, dusty day the Bisbee stage was held up and Billy Breakenridge and Dave Nagle, two of Behan's deputies, arrested for the crime Frank Stilwell and Pete Spence, both Clanton men. When the prisoners were arraigned and were about to be released on bond, Wyatt Earp as deputy United States marshal rearrested them for robbing the mails and caused them to be arraigned once more, seeing to it that they posted much heavier bonds before release. By the outlaw faction the deputy marshal's action was considered gratuitous. Threats came up to Wyatt Earp in Tombstone. Then a sudden incident in the town itself made the feeling more bitter. Into Tombstone the evening of October 27, 1880, rode a bunch of the rustlers. They were led by Curly Bill, Ike and Billy Clanton, Frank and Tom McLowery, and included, among others, Frank Patterson and Pony Deal. After a round of the saloons the whole crowd grew drunk and ugly, made threats to the passers-by on the streets and generally created a disturbance. Fred White, city marshal, asked Wyatt Earp to help him arrest the gang.

The two men closed in on Curly Bill, who was flourishing a pistol. White grasped the barrel of the weapon and wrestled for it. The gun roared and the marshal fell dying, shot through the bowels. Instantly Curly Bill was stretched senseless beside the man he had killed, pistol whipped by Wyatt Earp. From a dark alley other rustlers began to shoot, their bullets whimpering past the ears of Earp, who squatted on his heels by the fallen men, replying with his own six-shooter. Presently the outlaws retreated. Earp placed Curly Bill in the calaboose, saw White cared for, then went silently and grimly to arrest the rest of the gang. One by one he hunted down Frank Patterson, both McLowerys, Billy Clanton, and Pony Deal. Each in turn he treated the same way. Walking up he struck the outlaw a tremendous blow on the head with the barrel of his pistol, then had the senseless man carried to jail.

It was a spectacular display of individual daring, induced by cold fury at the shooting of Fred White, Earp's friend. Before he died, however, White exonerated Curly Bill, saying the shooting was accidental, due to the discharge of Brocius' pistol which had a hair trigger, while the two were struggling for possession of it. Thereupon the prisoners were freed with a reprimand from the judge. They rode back to their own bailiwicks at Galeyville, Charleston, and elsewhere down toward the border, cursing the Earps and crazy to pay them back.

Almost a year passed, however, before the Clantons, on October 25, 1881, once more rode into Tombstone and began to "licker up." In a saloon Ike Clanton and Doc Holliday met. Neither was armed, but they abused each other in picturesque language, and a fist fight was averted only because Virgil Earp pulled his friend away. The tubercular dentist would have been no match for the brawny cattle rustler. Ike was still drunk and aching for trouble. "Go sleep it off," Wyatt Earp sternly advised him. Clanton staggered away.

Early next morning the Earp brothers, who lived together, were awakened with the news that the Clanton clan was gathering. On a prominent street corner stood Ike, armed with a Winchester and proclaiming that he was either going to run the Earps out of town or "eliminate" them. He was joined presently by his brother Billy, the two McLowerys, and Billy Claiborne. They made a little knot of five dangerous gunmen, ready to have it out at last with the Earp brothers whom they all hated. Later, however, the Earps, coming down the street, found Ike Clanton alone, drunker than before, "buffaloed" him and took him to jail where the judge fined him twenty-five dollars. That further inflamed the rustlers. It would be tedious here to recount the threats, challenges and insults of the next few hours. The climax came when Sheriff Behan met Virgil Earp and told him, "Ike Clanton and his crowd are down at my corral making gun talk against you fellows. What are you going to do about it?"

A few minutes later R. J. Coleman, a miner, came into Hafford's saloon where the Earps were lounging. The Clantons and McLowerys, he said, had shifted their position to the O.K. corral, whence they sent by him a message: "Tell the Earps that we're waiting at the O.K. corral, and if they don't come down and fight it out we'll pick them off the street when they try to go home. If Wyatt Earp will leave town, we won't harm his brothers, but if he stays the whole outfit will have to come down and make its fight."

It was the formal cartel of the West. Without another word, although Captains W. B. Murray and J. L. Fronck, leaders of the newly organized vigilantes, offered to call out their men and disarm the outlaws, the Earp brothers started down the street toward the O.K. corral. On the way they were joined silently by their friend, Doc Holliday. The little, sunk-chested dentist carried a sawed-off shotgun.

They made a grim and threatening group—three tall, hard-visaged Earps, and their cadaverous, ghastly friend. On the way Sheriff Behan tried to stop them by telling them they all would be killed—there were five in the Clanton force to oppose the four in the Earp array. The four only strode on.

Into the O.K. corral they turned. The Clanton fighters were waiting in a sort of battle line. Shouts and hysterical warnings sounded up and down the street, drowned by the sudden shattering explosions of many guns fired at once. In sixty seconds it was over except for the haze of smoke drifting slowly across the corral. Morgan Earp was down with a bullet through his shoulder, and Virgil was staggering, shot through the leg. But Frank McLowery lay dead, Wyatt Earp's slug through his stomach and Morgan's through his skull. Tom McLowery was dead also from the double blast of Doc Holliday's riot gun. With a hole from Wyatt's six-shooter through both his hips, Billy Clanton was dying from hemorrhage. And Ike Clanton, who had boasted so loudly, with Claiborne, was running, gasping, hunting for safety down the street.

Few shorter and more decisive gun fights ever occurred in range history. After his wounded brothers received medical care, Wyatt Earp sternly sought out and arrested Ike Clanton and Claiborne and placed them in jail. Later in the day Sheriff Behan attempted to arrest the leader of the Earps for murder, but the mere glare of the deputy U. S. marshal changed his mind on that score, and a coroner's jury shortly afterward exonerated the brothers.

From the fight in the O.K. corral dated the decline of the rustlers of Cochise County. Virgil Earp recovered from his wound, only again to be wounded by an unknown assailant, weeks later. He recovered from this wound also. Morgan Earp was less fortunate. Five months after the showdown battle, he was shot dead from the darkness outside through a window of Bob Hatch's saloon in which he was playing billiards one evening.

Now it was a question of blood-payment for the Earps and in deadly earnest they ranged the country for their enemies. Within forty-eight hours Frank Stilwell was killed by Wyatt Earp near the railroad track in Tucson. Stilwell was one of the men arrested for the Bisbee stage holdup, the genesis of the feud between the Earps and the rustlers, and Wyatt believed he had fired the shot that killed Morgan.

Once more Sheriff Behan made a half-hearted attempt to arrest the tigerish leader of the Earps, but public opinion favored the brothers and other officials refused to cooperate. Warren Earp came riding into Tombstone from California to take Morgan's place, and to join the posse of proven men Wyatt was organizing. For a month that posse hunted through Cochise County. Indian Charlie, a half-breed hanger-on of the rustlers, was killed at the camp of Pete Spence. Before he died he definitely implicated Spence in the murder of Morgan Earp, for which Spence later served a prison sentence.

Everywhere as the relentless posse rode, rustlers, road agents, and other outlaws scattered. Word was out that it was death to be caught by the brothers, and John Ringo, Hank Swilling, Pony Deal, and Ike and Phin Clanton fled to Mexico. Curly Bill Brocius remained and dared to meet Wyatt Earp in a duel with guns. Curly Bill had the first shot, but then the deputy marshal cut him down with buckshot from a Wells-Fargo shotgun. Here and there an outlaw, more daring than his fellows, still clung to some isolated water hole, but the strength of the rustlers was broken.

Their campaign was over and the Earps departed from Arizona to Colorado, where they were still to encounter some legal troubles over the shootings in Tombstone, and eventually to win full vindication. How well they had done their work can be seen by looking over the subsequent record as well as the fate already established of the leaders of the once powerful outlaw bloc: Old Man Clanton and Billy Clanton were dead. So were the McLowerys, Frank and Tom. Ike and Phin Clanton were in Mexico, but Ike was killed soon afterward by Sheriff Commodore Owens, along with Pony Deal, and Phin went to the penitentiary for cattle stealing. Curly Bill Brocius was dead, together with Frank Stilwell, John Ringo, Jim Crane, Harry Head, and Billy Leonard. The rest were gone from the country. Law and order had come to Cochise County . . . after a fashion.

XXXV

SHEEP AND COWS DON'T MIX

There is a saying of the West that ought to be elaborated for the sake of strangers: sheep and cows don't mix. It is to some minds inexplicable that the one continuing, unending feud in the range country should be that between these two very similar forms of animal husbandry. Deaths in large numbers have occurred over it, and to this day in many sections of the cattle country the name "sheepherder" is a fighting insult. Yet for all this there is a natural explanation if you but consider the standards and ways of the time and country.

Almost from the first the animosity between cattleman and sheepman was cordial and mutual. In the beginning it probably had a racial angle. While Texas was developing into the first of the great cattle states, New Mexico and Arizona were devoted primarily to sheep. The cowpuncher of Texas, with due allowance for the numerous *vaqueros* who rode with him, was an American. By the same token the sheepherder, or *pastor*, of New Mexico was a Mexican or Indian. The two callings fostered the mutual distrust and dislike the races held for each other—a prejudice that spread over the West as the range was opened for both sheep and cattle and rivalry mounted for possession of the grazing lands.

But there were reasons even more fundamental. The cowboy was the pioneer of the range. He it was who fought the Indians, explored the new country, and occupied by right of discovery the great pastures. After a time the range became crowded with cattle, which was a situation bad enough, but then came other thousands of woolly, stupid, dirty yellowish-gray creatures to further compli-

cate the problem. When the sheep began to appear cowmen realized that free range was almost all taken. Farther
and farther into the West each year reached the barbed
wire fences. To the strenuous, wild riders of the horned
herds, the loss of their boundless horizons meant they must
change or abandon their chosen form of existence, and
naturally they resented it. Many of them blamed the sheepmen for it.

Between cowboy and sheepherder was also a physical
and mental difference. The cowboy exulted in living the
hardest life on the continent. His existence was a continuous stirring struggle against the savagery of Nature and he
was proud of his ability to survive it. He was swift, vigorous, reckless and dashing. On the other hand, the sheepherder lived a solitary life, often afoot, with no companions save his dog and his thousand or so sheep. Unlike cattle, sheep completely lack individuality, and their similarity makes them maddeningly monotonous. Dulled by the
loneliness and sameness of their existences sheepherders
sometimes become sullen, dreamy, or "queer." To the
cowboys they were a caste apart, ununderstandable, immeasurably inferior in the rudimentary social scale of the
primitive West.

But there was still another cause for hatred. It was based
on the difference in the grazing habits of the animals. Cattle are mildly gregarious only. Over the country they wander loosely, their method of grazing being to loop bunches
of grass into the mouth with the long and muscular tongue,
which means that they do not graze the range very short.
Conversely sheep move always in compact bunches, their
edged little hoofs cutting up the sod and pounding it hard.
Their sharp incisors and the split upper lip enable them to
bite off the grass to its roots and below if they desire.
When they say in the West that a certain range has been
"sheeped off," they mean a condition of destruction where
a once-smiling grassland has been sheared, hammered and

scarred into a virtual desert. It was formerly held, moreover, that sheep polluted a country in such a way that
cattle would have nothing to do with it. Drinking holes
were so trampled and defiled by the woollies that they
were unfit for other animals.

From the first, as economic differences were given an
edge of rancor by sheer prejudice, the cowboys took active steps to stop invasions by the sheep herds. Charles
Goodnight established a "deadline" as early as 1876, the
winter before he went down into the Palo Duro canyon.
He had trouble in the Canadian valley as has been recounted, and when he took his cattle the following spring
into the Staked Plains he made a "treaty" with the sheepmen that he would leave them the Canadian if they would
not cross the divide into the Palo Duro. That agreement
was kept for many years.

Last of the real sheep wars was fought in the magnificent Ten Sleep country of Wyoming, where gigantic maroon cliffs dwarf into minuteness the tiny cabins of the few settlers who dwell in the midst of the vastness. There long had been trouble between cattle and sheep interests in Wyoming, but it came to a final climax in 1909.

One after another, ranges long held exclusively for cattle were invaded by the noisy, destructive, ill-smelling flocks. Of course the cowboys swept down upon them at once. Usually the sheepherders were quite helpless—a man or two and a few dogs, against a squad of a dozen or so armed horsemen. Sometimes a band of range riders would stand off from a sheep herd and shoot into it with Winchesters until ammunition was exhausted and scores of dead, woolly forms lay around. If the herders objected the bullets flew in their direction. Many a humble Mexican, Basque or American sheepherder paid with his life for trying to protect his flock—but it should not be thought that the shooting was, by any means, all on one side. A good many cowboys, too, have been knocked off their horses from a distance by "scab herders" proficient with Winchesters, and left to the coyotes out under the wide Western skies. Another favorite method of sheep extermination was by driving a herd over a cliff. Immense, thousand-foot precipices are common in the Ten Sleep country. It was necessary only to start the leaders over and the rest of the stupid bleaters followed. Or saltpeter could be spread on the range and the sheep ate it and died.

In spite of the barbarities practiced against them, however, the sheep, like an uncontrollable, viscous flood, still poured into the range. Down in the Sweetwater country, east of the Wind River Mountains, the first deadlines of Wyoming were established. Later the practice was adopted all over the state. Imaginary lines were drawn across the country, and the sheepmen notified that if their flocks passed those lines, sheep and herders alike stood in peril of death. Nor were these idle threats. When the arbitrary boundaries were violated, the cowmen struck—quickly, in some cases with shocking brutality.

There is the case of Louis A. Gantz, whose herd of sheep was destroyed August 24, 1905. Gantz' sheep camp, about forty miles from the little town of Basin, was attacked in the night and about four thousand out of seven thousand sheep shot or clubbed to death. A team of horses, a supply of grain, provisions, and a wagon and camp equipment were destroyed. The foray cost Gantz about $40,000, but he made no attempt to prosecute his attackers. Too well he knew the temper of the Wyoming courts at that period toward such as himself.

Atrocity in the end, however, reached a limit that aroused even the prejudiced public opinion of Wyoming. In the heat of that indignation the war between the cowmen and the sheepmen burned out. Joe Allemand, a sheepman from Casper, made his camp on No Water Creek, in the Ten Sleep country, April 3, 1909. With him were his camp mover, Joseph Emge, and a herder, Jules Lazier. The three men were asleep by their fire, when a band of killers crept up on them, murdered all three, piled the bodies on the camp dunnage, and burned the heap. Then the marauders turned on the sheep and destroyed them by hundreds.

Too brutal even for a calloused country was this crime and the Wyoming Wool Growers' Association took courage to offer large rewards for the arrest of the killers. On evidence gathered by range detectives, a grand jury at Basin early in May filed true bills against seven Ten Sleep country cowboys, George Sabin, Herbert L. Brink, Milton Alexander, Ed Eaton, Tom Dixon, Charles Faris, and William Keyes. Faris and Keyes turned state's evidence. The others were convicted of crimes ranging from murder to arson and were sent to the state penitentiary. The Ten Sleep war abruptly came to an end.

The Wyoming troubles were nothing, however, compared to the Graham-Tewksbury feud of Arizona, known in the Southwest as the Tonto Basin war. It began in the summer of 1887 and continued until the last of the Grahams was killed by the last of the Tewksburys in 1892. No feud of such unexampled ferocity is known elsewhere in the West.

Wide and smiling, Pleasant Valley lies under the frowning immensity of the Mogollon range of Arizona. It was occupied by ranchmen as early as 1880, among the first being John Stinson. By 1887 several ranches were in the valley, including Stinson's, the Aztec Land & Cattle Company—familiarly known as the Hashknife—and the Graham ranch. Sheep had appeared in the ranges to the north, but Pleasant Valley cowmen designated the Mogollon peaks as the sheep deadline and the sheepmen knew and respected that ukase.

On one of the smaller ranches lived the Tewksburys, a grim, wry strain of frontiersmen. The patriarch of the clan was old John D. Tewksbury, Sr., a Boston Yankee who had made the gold rush to California in '49. By an Indian wife he had three sons, Ed, John, Jr., and Jim—as harsh and even more downright than himself. The Indian wife died and the elder Tewksbury, after he moved to Arizona, married an Englishwoman, Lydia Crigler Shultes. All male members of the family were magnificent revolver and rifle shots; all, perhaps because of their Indian blood, were adepts at scouting, trailing and bushwhacking. They all had, moreover, a taciturn scorn of compunction against taking human life which was to give them a sinister reputation.

The veteran rancher, John Stinson, earliest comer to Pleasant Valley, wished to enlarge his holdings, and employed to work in his interests the Grahams—Tom, John and Billy—typical cowmen, boastful, courageous, and somewhat reckless. Thus the two sides were aligned in the vendetta that was about to break out—the Grahams representing the cowmen and the Tewksburys the sheepmen.

It is impossible to arrive at the true facts concerning the start of the feud. Thirty years ago William McLeod Raine made a competent effort to do so, but failed to unravel all the story's devious strands. More recently Earle R. Forrest devoted years of investigation to the subject and his book *Arizona's Dark and Bloody Ground* is the best authority on the Tonto Basin war. Yet even he was unable to push aside the whole curtain which shrouds the events that took place. Certain it is, however, that there was an early animosity between the Graham and Tewsbury families. There is a story that one of the Graham men showed attention to one of the Tewksbury women, occasioning jealousy. It is known also that charges of rustling were made against each other by both sides, and a possible motive for the quarrel is contained in the report that John Stinson tried to buy out the Tewksburys through the Grahams and anger flared when the half-Indian family rejected the offer.

Not, however, until sheep entered the controversy did the ill-feeling result in blood-letting. That was early in 1887. A cowboy, riding in the northern edge of the valley, had seen that which made his eyes bulge and he carried the news as fast as his pony could travel: the Tewksburys were driving sheep over the rim of the Mogollons!

Cowmen gathered in knots and discussed it. This was alarming news—fighting news. At first nobody believed it. But other cowboys soon verified the report. There, slowly pouring over the landscape, came the yellow-gray flood of hated animals.

Two brothers, P. P. and W. A. Daggs, who long had looked enviously at the rich grass of Pleasant Valley, owned the sheep. Partly to spite their enemies the Grahams and the other cowmen in the valley, and partly for the sake of a share in the flock increase promised by the sheep growers, the Tewksburys had contracted to give the protection of their guns to the herds that were brought into the valley. It was a combination of insult and injury to every cowman—after wresting the valley from the loping Apaches, to have it taken over by blatting sheep!

Hashknife cowboys shortly reported the sheep were causing their cattle to withdraw from the water holes. A formal warning sent to the sheepmen was ignored. A few days later one of the Daggs herders was killed. He was a Navajo Indian who has not even the dignity of a remembered name, but his courage is worthy of commemoration. Although he knew men were seeking his life, with scrupulous loyalty to his employers he stayed on his job—and died. That was in February, 1887.

Soon after a band of cowboys chased three sheepherders up into a rocky mesa and held them there under their rifles while in full view their sheep were slaughtered. By the end of the month the last straggling remnants of the flocks were driven back across the Mogollons.

This was bitter cordial to the Tewksburys. They had expected to profit from the sheep venture. Their pride rankled also at the realization that in their first tilt with the Grahams the latter had won. Almost at once sinister occurrences were noted. Mark Blevans, a veteran cattleman, disappeared in July in a manner so mysterious that it has never been explained. The Graham faction immediately charged the Tewksburys with his murder. From that day the old rancher's four sons, Andy (known also as Andy Cooper), Hampton, Charles and Sam Houston Blevans, a fierce group with a more than ugly reputation, became the foremost advocates of violent action. By this time the entire valley was in the grip of the feud. There could be no neutral ground. It was a case of siding with one element or the other—or leaving the country. Most of the cattlemen backed the Grahams. The Tewksbury partisans were fewer, but they more than made up for it by their lethal skill in ambush and concealment, and their deadliness with their guns.

Hampton Blevans, blind mad with rage over his father's death, started the open hostilities. In spite of efforts by Tom Graham, a clean-cut cowman, to avoid bloodshed, on August 10 Hampton Blevans, accompanied by Tom Tucker, John Paine, Bob Gillespie, Bob Carrington and three Hashknife cowboys, rode up Cherry Creek looking for the Tewksburys. That hunt was to be fatal to some of the hunters.

The Tewksburys were found at a place known as the old Middleton ranch. At this stage of the feud most of the antagonists did not know one another and the cowboys seem to have spent some time in conversation with the dark loitering men, suspicious but not sure these were the people they sought.

What set off the explosion nobody really knows. A gun roared, a man spun out of his saddle. Horses reared and snorted. For thirty red-hot seconds a furious cachinnation of shots shivered the air. Then the surviving cowboys rode for safety. Two of them were dead—Hampton Blevans and John Paine. A third was shot horribly through the lungs and seemed certain to die, although he later recovered, almost miraculously. That was Tucker. Bob Gillespie and Bob Carrington also were wounded, but less seriously. Not a sheepman was hurt.

The cattle contingent had been stunningly defeated, but the Tewksburys were realistic and knew it was time to hunt cover. All up and down the valley the cowmen would be gathering their forces to return to the attack. Sure enough, the sheepmen, "forted" on a high butte, were discovered at last by part of the cowboys and for one whole long day the Tewksburys were besieged there, the Winchesters crackling desultorily as each side vainly sought to find a target. When night came the Tewksburys were suffering from thirst. One of the brothers, Jim, climbed down to a spring that flowed near and filled two canteens with water. With these slung over his shoulder, and his rifle in his hands, he had started to return, when suddenly from the tip of the mesa where one of his brothers stood guard, came a cry of warning. Without glancing behind, the water carrier pointed his rifle, muzzle backward, over his shoulder, and pulled the trigger. The shot was amazingly lucky. It brought down an assailant sneaking up behind, the bullet piercing the thigh of a cowboy whose name is forgotten, but who bled to death while the remorseless Tewksbury rifles held back his friends from aiding him. After that the cattlemen once more withdrew.

Word now reached the countryside that a feud was raging in Pleasant Valley that had already cost five lives. Sheriff William Mulvenon of Yavapai County secured warrants for the Tewksburys, charging them with murder, but his "invasion" with a large posse was fruitless. None of the lurking sheepmen could be found and no arrests were made.

Almost immediately thereafter tragedy brought directly home to the Graham family the seriousness of the war. Billy Graham, a carefree, handsome boy of twenty-two, having been to a dance at Phoenix, was riding home on August 17, when he was ambushed and mortally wounded. Years later it was learned the slayer was James D. Houck, a sheepman and Tewksbury supporter, who had somehow secured a deputy sheriff's badge, and used it to give him authority to bushwhack the country. The horse of the boy carried him to his brother's ranch where he died.

Bitter with grief and rage at Billy's death, the Grahams gathered their cowboys and once more hunted the deadly Tewksburys up and down the valley. At last the sheep crowd was located at the Tewksbury ranch. Seven men and three women were at the tumble-down cabin, but when the Graham party arrived two of the men, John Tewksbury and Ed Jacobs, had left to bring up some horses. It was early in the morning of September 2. No warning came to Tewksbury and Jacobs, and they blundered right into the trap set by the cowmen. Andy Blevans, the implacable, slew both sheepmen. The intense bitter-

ness of the hatred in the feud made itself evident in the
succeeding events. In plain sight of the house where by
now the rest of the Tewksbury clan lay fighting desperately in the defensive, lay the bodies of the two slaughtered men all day. While the bullets sang back and forth
a drove of half-wild hogs filed out of the underbrush and
began rooting and tearing at the bodies. No move was
made by the cowmen to drive the swine from their revolt-
ing repast. To their mind being eaten by hogs was a fit fate
for a sheepman.

The men in the besieged Tewksbury house did not dare
to try to reach the bodies. It would have meant death. But
after a time John Tewksbury's widow could stand it no
longer. Out from the house she marched desperately, a
shovel over her shoulder. She relied, and rightly, upon the
tradition of the West not to harm a woman. Not a shot
was fired as she went to where the dead men lay, drove
off the hogs, dug two shallow graves, rolled the bodies in
them, and covered them with earth. Then she returned to
the house and a short time later the cowboys mounted and
departed. Throughout the war the Tewksburys generally
outlasted their more mercurial foes.

Unexpected assistance from without struck heavily in
favor of the sheepmen two days later. The Blevans family,
fiercest of the Graham supporters, had moved their resi-
dence to Holbrook, northeast of the valley. This was a mistake because Sheriff Commodore Perry Owens, of Apache
County, had a warrant for Andy Blevans, charging him
with stealing cattle from the Navajos. The sheriff attempted to serve his warrant, but the whole Blevans family resisted the arrest with six-shooters. Standing outside
the house, with his enemies inside it shooting at him, Sheriff
Owens killed Andy and Sam Houston Blevans, together
with Mose Roberts, a brother-in-law, and badly wounded
John Blevans, who was later sent to prison. In one brief
burst of firing the Blevans family was all but wiped out
and the Grahams lost their most redoubtable gunmen.

Sheriff Owens, who gave this remarkable exhibition of six-shooter efficiency, was the same man who completed Wyatt Earp's work in Cochise County by killing Ike Clanton and Pony Deal.

This loss, which occurred September 4, did not prevent the Graham faction from making another sudden attack on the Tewksburys, September 7, when the sheepmen were camped on Cherry Creek. It was an ill-advised move. Right through the camp the cowmen rode early in the morning, but the Tewksburys simply lay in their blankets and shot back. Harry Middleton was killed and Joe Ellenwood was wounded among the assailants and the cowmen once more were driven off.

The luck was breaking against the Grahams. Four days later, on September 11, Sheriff Mulvenon made his second invasion of Pleasant Valley. This time he took along a strong force of deputies, with the intention of hunting down both sides of feudists. He was met by Houck, the slayer of Billy Graham, who possessed, as has been noted, a deputy's badge. Attaching himself to the posse, Houck convinced the sheriff that the Graham crowd was entirely in the wrong, and acting on this belief Mulvenon set a trap. On September 21 the posse ambushed John Graham and Charles Blevans, sole survivor of the Blevans clan, and killed them both. After that the sheriff arrested men of both sides, but when the cases came to court no witnesses could be found willing to testify and the criminal charges were dropped.

For months the vendetta dragged along. To attempt to recount every ambuscade, every surprise attack, every killing, as they occurred, would be wearying. Several men not connected with either side were "dry gulched" on mere suspicion. One of the sickening episodes was the lynching of three young men, James Scott, James Stott, and Billy Wilson, by a gang of Tewksbury supporters allegedly led by Houck. Earle R. Forrest has compiled a list of twenty-seven men, of whom six were never identified, who died in the Tonto Basin war. Of these only four were Tewksbury partisans, twelve were Graham men, and the rest were strangers or innocent parties, killed because they blundered into dangerous territory.

The eldest and last of the Grahams, Tom, was murdered August 2, 1892, while he was preparing, at the pleading of his wife, to leave Pleasant Valley and its feud. Ed Tewksbury, last of that clan, was convicted of this murder, but obtained a new trial and eventually the case against him was dropped. Thus, of the Grahams, all were dead. Of the Blevanses all were dead but one and he was in prison. Of the Tewksburys only Ed survived, Jim having died of illness during the vendetta.

And the upshot? As elsewhere in the West where sheep and cattle interests clashed the sheepmen eventually won. Why? Because of their dogged, ceaseless pressure. The hard-riding, impetuous cowboys finally had to give way before the slow relentless advance.

Perhaps this result was helped, not only in Pleasant Valley but elsewhere, by the discovery that one of the chief arguments of the cattlemen against the presence of sheep on the range were in error. Properly pastured, sheep do not devastate the range, nor do they befoul water and "leave an odor that cattle will not tolerate," as formerly charged. There is, moreover, as much profit in growing sheep as in cattle—or more. In this more enlightened state of knowledge the old deadlines were removed. Sheep were permitted to enter the range, and the hatreds died down in New Mexico, Arizona, Colorado, Wyoming and Montana. Today many ranchers grow sheep and cattle on the same range, owning both.

However, traces of the sheep and cattle wars can still be detected in the West. A "sheepherder" or "scab herder" remains a man without honor or social standing in the range country. And throughout much of the West it is not accounted polite to invite a man to eat mutton. The wild, flaming cliffs of the Ten Sleep country and the pines of the Mogollons, as well as the sage and cactus of the deserts, hold long-forgotten secrets of a fierce and murderous period. Occasionally old, bleached skeletons are found today in that vast country. There is no way to identify them. But the discoverers, giving the bones decent burial, say to one another, "Poor devil, he was sheepy."

XXXVI

CATTLE WAR IN THE ROYAL MANNER

A score or so of years ago you approached the Bell ranch, whose brand, the Dinner Bell, festooned the flanks of some thirty thousand cattle, by a road both devious and long among the mesas and foothills north of Tucumcari, New Mexico. Eventually you reached the front gate of the ranch—if your car held out over those trails—and then there was another fifteen miles, over adobe flats, up canyons and arroyos, and around and over hills and knobs, before you reached the front door of the headquarters house.

All this was Western and typical of the cow country, and you were accustomed to it. But once you stepped into the low, rambling adobe ranch house, with its patio and garden, the whole atmosphere changed—you were transported in a second from raw frontier America into cultured old England. A spot of tea awaited you, if you desired it. Except for the deep tan of his features the host might have stepped off Piccadilly, his jacket a Norfolk tweed, and his legs were encased, not in chaps, but in meticulously tailored riding breeches and polished boots. His accent was Oxford and his manners Mayfair. When dinner was served—at the summons of a butler—you entered the dining room in tails and a white tie if you were a gentleman; décolleté, if you were a lady. The table was set with fragile china and sparkling glass, the meal was served in proper courses, and at the end there were cigars for the gentlemen and well-bred drawing room conversation for all.

The manager of the Bell ranch who would have been your host as described was Charles O'Donel, a former officer in the British army, with a fighting record in a swank regiment in the Boer War. He loved the niceties and the luxuries of the British gentleman's existence, but that did not prevent his being a hard-working and efficient, if somewhat autocratic, cattleman.

There were many like O'Donel in the West—Moreton and Richard Frewen of Wyoming, whose log home ranch was like a royal hunting lodge, Sir Horace Plunkett, the Marquis de Mores, and other aristocrats of the range, both foreign and American, who lived on a scale and in a style that made the dusty riders of cattle land look askance and utter dry and wry witticisms to one another. A caste was being created in the West, which always had gone on the principle that one man was as good as another, regardless of the number of cows he owned or the name he bore. And out of it grew conflict. As the Lincoln County war was a struggle between rival interests, the Cochise County war between outlawry and law, the Tonto Basin war between cowmen and sheepmen, so the Johnson County war was a contest between the cattle kings and the small cattlemen. It was the last stand of the great royal houses of cattle land against the encroachment of the lesser settlers and was conducted for a time at least with all the pomp and circumstance suitable to a baronial war waged on usurping yeomen.

In Northern Wyoming the Johnson County war is a topic of discussion still. First and last it involved several hundred fighting men—the sun-baked, lean riders of the illimitable high plains and foothills country. Few were killed, comparatively, but this was no reflection on the deadly intent or shooting ability of the combatants. Rather it is a tribute to the Indian skill with which both sides took advantage of protection. It was a class war, with both sides sincere in the belief that they were right. The cattle kings called the small ranchers rustlers, and charged them with stealing cattle. The small ranchers called the cattle kings murderers and invaders and accused them of holding illegally the best grazing lands. But in the struggle the cattle barons called themselves "regulators," and the small ranchers referred to themselves as "settlers."

The genesis of the Johnson County cattle war, concerning which there are still traces of bitterness in Wyoming after the lapse of half a century, goes back to the days when the first trail herds came up the Powder River from Texas right after Sitting Bull's Sioux finally were chased on to their reservations. The first herds entering Wyoming were owned by wealthy men—"big operators" they were called in the West. Later, as we have seen, other large investors, including many British capitalists, came to the state. They liked the romance of the range, but in many cases had small understanding of the cattle business and less of the cattle people. Frequently big owners did not even live on their ranches, preferring the society of Denver or Kansas City, or at least of Cheyenne, to the lonesomeness of the plains—except for occasional visits made mandatory by business.

An antipathy which was wholly natural grew up between these cattle barons and the men who had small holdings and who lived and worked with their few head of stock winter and summer. Especially was this true when the small operators with their limited capital learned that the big outfits were holding hundreds of thousands of acres of grazing land to which they had no legal right, but which they maintained for their herds by force, with straight-shooting, heavily armed men in the line houses.

So great grew the bitterness of feeling that to prey on the opposing side came to be regarded as semi-ethical. All the political advantage was with the cattle barons whose money furnished them a mighty leverage on the state government. But the small ranchers were right on the ground, and their individual activity made up for the larger political influence of the great cattle firms. Calves that did not belong to them were branded by nesters. To railroad construction camps and reservations were driven steers with blotted brands, and the money paid for them did not reach the legitimate owners. Nor were the nesters and small ranchers alone in swinging the long rope and using the running iron. Some of the biggest cattle outfits applied brands to calves and even adult cattle owned by others. It was simply a case of divergence of feeling and difference of aims so wide that anything was considered fair that operated to the disadvantage of the opposing party.

To the big cattlemen the situation was intolerable and eventually they began sending cattle detectives into the country—men who, with tremendous nerve, filtered into the ranks of nesters and rustlers at the hourly risk of their lives, posing as cowboys, small ranchers, cooks, or even trappers, to learn the identity of the men who were stealing the cattle of their employers. Rewards were offered for the arrest and conviction of stock thieves, but though many men were arrested, and there appeared, sometimes, to be sufficient evidence to convict, prisoner after prisoner was acquitted by juries patently in sympathy with the small ranchers. In Johnson County, particularly, the political officials were elected by nesters and little cattlemen, and there was nothing the big owners, who elsewhere were in control, could legally do about this one stubborn county.

Only the appeal to force remained—but it was easy to invoke in the direct-action West. So it came about that early in 1892 a group of men met by prearrangement in a closed room of the exclusive Cheyenne Club. They were men of evident prosperity, yet with the ineffable something that stamped them as cattlemen. One who knew the cow country would instantly have recognized the gathering as a meeting of notables in the Wyoming cattle industry. Close-mouthed, determined, autocratic, they formed that day a secret "Regulator's Association," and laid plans which in this modern time sound almost incredibly sinister. The proposal was simply to hire a band of the deadliest gun-fighters in the West, invade Johnson County, hunt down, and kill like animals the worst offenders among the alleged rustlers. It was a cold, brutal, callous plan, but the cattle barons had reached the point where they believed any measures which would attain results were justified, no matter how extreme. More than seventy names were on the list of the proscribed gathered by the cattle detectives.

Major Frank E. Wolcott, a short, squat ex-army officer, was selected to command the "invasion," and the regulators, consisting of twenty-four six-shooter virtuosos imported especially from Texas, together with enough stockmen, detectives, ranch foremen, and trusted employees to raise the total to fifty-four, left Cheyenne on a special train consisting of coaches for the men and stock cars for the horses and saddles. In the gray of a blizzardy morning, April 7, 1892, they disembarked at Casper and at once rode northward toward the vast primitive country of which Buffalo was the natural capital.

A few early risers saw the party stealthily leave its train. There were then in Casper, Douglas, and many other towns, men who knew they might be marked for death by the sinister column. Rapidly the alarm spread, and everywhere Wyoming communities began to arm hysterically for defense, because nobody knew in what direction the blow would be struck. Only huge, remote Johnson County, where the communications were scanty, was ignorant of the deadly threat creeping upon it.

Riding rapidly northward, the regulators forced every man they met to turn back and accompany them for hours, lest he should send out word of the direction of their march. At the Tisdale ranch on the border of Natrona and Johnson Counties, the "army" halted, to rest and eat. The owner of this ranch was Bob Tisdale, a member of the regulators there present. Incidentally a J. A. Tisdale, who had his outfit not far away, had been killed the previous Christmas by an unknown assassin from ambush— why, nobody could say, although it was assumed that he knew too much for his health about rustling activities. He was returning from Buffalo where he had done some

Christmas shopping, and was found lying among the toys he had purchased for his children, which were stained red with his blood. Ranger Jones, a hard-working, honest homesteader, and Tom Waggoner, a small rancher, had also been killed mysteriously—Jones ambushed by someone hiding under a bridge at Muddy Creek, and Waggoner taken from his family and left hanging to a tree in what is now known as "Dead Man's Canyon." These three killings were vehemently charged by each side against the other. Even today the mystery of them is unsolved, but that they were connected with the range trouble is certain.

Resting at the Tisdale ranch, the regulators were on the very edge of hostile country. Shortly a scout, Mike Shonsy, foreman of the Western Union Beef Company, rode in with information that there were rustlers at the K C ranch on the Powder River.

The little settlement of Kay Cee, Wyoming, drowses today near the site of the old K C ranch. None of the original ranch buildings is standing now, and there is almost no trace of the old corrals and other structures. But in Kay Cee and its neighborhood are old-timers who could tell a great deal about the happenings there if they would speak —which they never do.

Before daylight, April 9, the regulators quietly surrounded the K C ranch buildings. It was a small, run-down outfit, formerly an outside camp of Moreton Frewen's, but had for some time been used as a hub of operations by Nathan D. Champion, against whom the cattle detectives had considerable evidence that he was a rustler, while it was known that he was a leader and organizer of the settlers' opposition to the big companies. One attempt had been made to arrest Champion, a very brave man and a dangerous fighter, but on that occasion he drove away the party that came after him, wounding two. On this day, however, it was to be a different story.

Shortly after dawn an old trapper named Bill Walker,

who had spent the night at the ranch house, left the house and walked down past the only other building on the place, an old barn, with a bucket to obtain water from the water hole about seventy-five yards away. When he turned the corner of the barn out of sight of the house, he found himself looking into the muzzles of half-a-dozen Winchesters, and kept his silence, trembling, at the gruff, low command from the men who held the weapons. Walker was taken to the rear of the regulators' line for questioning. Badly frightened, he readily told them three men were still in the cabin—Ben Jones, Walker's trapping companion, Champion, and his partner, Nick Rae. The last two were the ones wanted by the regulators. In a few minutes Jones followed his partner down the path and was silently captured as had been Walker. Then the men with the Winchesters waited tensely for the appearance of the other occupants of the cabin.

Nick Rae was first to step out. As he stood looking about in the early daylight, a vicious volley crackled from a score of hiding places. Screaming the settler fell, shot through the head and body. Blindly he tried to drag himself back to the cabin, but more bullets thudded about him and into him. He reached the door. Suddenly the portal was flung open and Nate Champion, the bravest man and best revolver shot in Johnson County, leaped forth. At the roar of his six-shooter, D. Brooke, the Texas Kid, one of the hired gunmen with the regulators' force, doubled up, wounded. A moment later and Champion, unharmed by the return fire of bullets that rattled about him, had dragged the wounded Rae within and slammed the door.

The Texas Kid was taken to the rear and his injured arm treated, while the fight settled down to a siege. All day, rifles cracked and bullets smashed through the windows and doors of the cabin. Single-handed, Champion defended himself ably, firing now here, now there, being at all points of the house, it seemed, at once. Two more regulators were wounded and still the cabin held out.

About 3 o'clock that afternoon Black Jack Flagg, a settler, rode by on the road a few hundred yards away, with
his stepson, seventeen years old. The boy was driving a
team with the bedless running gear of a wagon, while
Flagg was mounted and riding beside him. On these way-
farers the regulators, with surprising stupidity, opened fire
at a distance, apparently hoping to prevent them from
spreading news of what was going forward. But the shooting only insured the spread of the news. Flagg and his stepson cut the traces of the team, the boy mounted one of
the horses, and the two of them galloped away, thoroughly
scared and full of information, but unwounded.

Now, the regulators knew, they must act quickly. There
was one way to oust Champion from the cabin before help
could reach him—they must burn him out. From the road
they conveyed the wagon gears deserted by Flagg, took
them to the barn, and there piled them with kindling, pine
knots, and hay. Using this creation as a screen from Champion's bullets, several men pushed it up against the beleaguered cabin and set it on fire. The flames quickly
caught on the log building, spread to the roof, and lapped
greedily; within a few minutes that whole side of the
shanty was ablaze. Higher and higher roared the leaping
fire. Inside it the smoke and heat became too great to en-
dure. Bootless and hatless, Champion at last dashed out
with his Winchester and revolver, to fall, game to the last,
under the bullets from a score of waiting rifles.

The most curious phase of the episode was that Champion kept a diary of the hours of his siege. Soaked with
blood and with a bullet hole through it, the notebook was
found in his pocket, and remains one of the most remarkable documents ever produced by the West. Apparently
with the full understanding that he must die, Champion
coolly jotted down his final impressions. The entry, written in pencil, under the date of April 9, is as follows:

Me and Nick was getting breakfast when the attack took
place. Two men were here with us—Bill Jones and another

man. The old man went after water and did not come back. His friend went out to see what was the matter and he did not come back. Nick started out and I told him to look out, that I thought there was someone at the stable and would not let them come back. Nick is shot but not dead yet. He is awful sick. I must go now and wait on him.

It is about two hours since the first shot. Nick is still alive. They are still shooting and are all around the house. Boys, there is bullets coming in like hail. Them fellows is in such shape I can't get back at them. They are shooting from the stable and river and back of the house. Nick is dead. He died about 9 o'clock. I see a smoke down at the stable. I think they have fired it. I don't think they intend to let me get away this time.

It is now about noon. There is someone at the stable yet. They are throwing a rope out at the door and dragging it back. I guess it is to draw me out. Boys, don't know what they have done with them two fellows that stayed here last night. Boys, I feel pretty lonesome just now. I wish there was someone here with me, so we could watch all sides at once. They may fool around until I get a good shot before they leave.

It's about 3 o'clock now. There was a man in a buckboard and one on horseback just passed. [This was Flagg and his stepson. Champion was mistaken as to the type of equipage.] They fired on them as they went by. I don't know if they killed them or not. I seen lots of men come out on horses on the other side of the river and take after them. I shot at the men in the stable just now; don't know if I got any or not. I must go and look again.

It don't look as if there is much show of my getting away. I see twelve or fifteen men. One looks like [name was scratched out, probably by one of the regulators after the diary was found]. I don't know whether it is or not. I hope they did not catch them fellows that run over the bridge toward Smith's. They are shooting at the house now. If I had a pair of glasses [binoculars] I believe I would know some of these men. They are coming back. I've got to look out.

Well, they have just got through shelling the house like hail. I hear them splitting wood. I guess they are going to fire the house. I think I will make a break when night comes, if alive. Shooting again. I think they will fire the house this time. It's not night yet. The house is all fired. Goodbye, boys, if I never see you again.

NATHAN D. CHAMPION.

If that is not a dramatic document the West never produced one. The man had a natural gift for expression. In his brief scribble he lives before you, with his solicitude for his dying friend, his rising and falling hopes, his analysis of his enemies' tactics. Observe that no thought of surrender or of mercy entered his head. He was probably guilty of the things charged by the regulators, and knew why they were there and what his fate was sure to be. But he did not for one instant lose his iron courage.

There was no sentimentality among the regulators. They left Nate Champion's dead body stretched on the ground where it had fallen, and someone placed on its bosom a rudely lettered placard: CATTLE THIEVES, BEWARE! Then the march of the "army" was continued toward Buffalo, with the ruins of the death cabin smoking behind.

Meantime, however, Flagg and his stepson had carried word as fast as their horses could travel to Buffalo, and all Johnson County was humming like a hornets' nest. In Buffalo, Sheriff Red Angus, who was a settlers' man, was swearing in deputies as fast as he could pass out stars to men who crowded up to volunteer. By the time the regulators reached the T A ranch on Crazy Woman Creek, just fourteen miles south of Buffalo, Phil Dufran, a cattle detective who had been scouting far out in front, galloped back to the column at the dead run, brought his foaming horse to a jolting stop before Major Wolcott and shouted out a stunning piece of news. Not less than a hundred armed men were riding south toward them from Buffalo.

Wolcott hastily called a council. If Dufran was correct —and they knew the veteran detective too well to doubt him—the situation was very serious. With inferior numbers

there was only one recourse open to them: they must for-
tify the T A ranch where they now were and prepare to
withstand a siege, hoping news of it would reach their
friends outside and bring rescue.

Spurred by common alarm, every man in the little expe-
ditionary force went feverishly to work. Trenches were
dug, rough barricades were built of heavy timbers recently
brought to the site to construct a new house, and Major
Wolcott coolly told off detachments to guard different
sides. In comparatively recent years you could still trace
the lines of the old trenches about the ranch houses.

Barely was the fort finished and the men posted, when
the enemy appeared, crowning the rises to the north. At
long range many rifles began to bark. Nobody was
wounded, but the settlers sent their horses to the rear and
began creeping up like Sioux, so that the situation grew
more and more grave. Hour by hour the firing grew heavier as the forces of the besiegers were augmented by constantly arriving newcomers. The regulators, who had so
cold-bloodedly invested Nate Champion in his cabin,
found themselves in a trap equally relentless. Closer and
closer drew the lines of the settlers. They were pushing
forward bales of hay behind which they dug rifle pits.
Three leaders directed them—Sheriff Angus; a Methodist
evangelistic preacher, the Rev. M. A. Rader; and Arapaho
Brown, an old scout and Indian fighter.

Inside of Wolcott's fort the Texas gunmen swore
grimly. They were accustomed to the thought of death
and were not panic-stricken now. The major, too, was
stern and resolute, an impressive figure for all his short
stature and strutting walk. Some of the less experienced
cattlemen, however, looked pale. Like Nate Champion
they may have been saying: "We don't think they intend
to let us get away this time."

And now, slowly, toward the ranch house which was
the chief bastion of the defense, moved a large smoking
object. It was a movable fire screen, built on the trucks of

a wagon similar to that which had been used against Champion. With a vengeance the regulators were learning how it felt to have their own tactics used against them. They were to be burned out and shot as they fled.

But the tragedy never was enacted. At the last moment, as if the scene were out of a melodrama rehearsed for the purpose, a bugle call came over the hills. Then against the magnificent background of the towering Big Horn peaks, a cloud of dust was seen coming down the road from Buffalo, and this presently resolved itself into three troops of trotting cavalry. Just in time the United States army had decided to take a hand in the Johnson County war.

That ended the fighting at the T A ranch, although several men were later killed in small individual combats out on the plains. Glad to surrender to the soldiers were the regulators, who had started out with such grandiloquent ideas. Later they were all brought into court, but they never were tried. By mutual consent, after continuances, the Johnson County matter was dropped. A better understanding had come up between all classes. The cattle kings at last were convinced that not even war carried on in a high and royal manner would enable them any longer to continue arbitrary domination of the range, and the small ranchers also were willing to stop abuses.

XXXVII

DUST STORMS AND DESOLATION

The year 1893 is commonly set as marking the end of the frontier, because in that year the government announced that it had no more free land of good grade to give away to settlers. The end of the open range came five years later, in 1898, for by then the last of it was stocked.

To the cow country both dates are significant. They betoken great changes that took place, changes much more than revolutionary, because they practically obliterated a whole way of life and set a new one in its stead. The plow brought the little cabin to the range country and the cabin brought the women. In the cow empire women had been few. Homes were scarce and a nomadic life has never been to the liking of femininity. Such women as did appear in the squalid little cow towns, too frequently were not of a character to be admired, although to be sure the West had probably more than its share, in proportion, of beautiful and fine women, with courage, and imagination, and loyalty—the wives and daughters of business men in the towns or of ranchmen on the range. But the cattle civilization essentially was a male civilization. Because of its freedom from conventions it established customs and ideas of its own, wild, direct, and somewhat violent.

Order and law came slowly and progressed by hitches. Some of the frontier jurisprudence was characterized by informality, to say the least. Judge Roy Bean, the justice of peace who called himself "the law west of the Pecos," was called upon to rule on the guilt of a cowboy who had killed a Chinaman. Having given the question deep consideration, his honor said that he had failed to find, in a long and careful perusal of the Constitution of the United

States, anything that deterred a white man from killing a Chinaman, therefore the cowboy should go free. A search was then made of the defunct celestial, which revealed that he was in possession of a revolver and a twenty dollar gold piece. This put a new light on the case. Judge Bean promptly reconvened court and fined the dead Chinaman twenty dollars for carrying concealed weapons—the gold piece being appropriated by the court in payment of the fine.

Mose Hays, owner of the Springer ranch, was called to Mobeetie as a witness in a case against a government butcher who had been slaughtering range cattle without paying for them. Court was presided over by Judge Frank Willis, a fine old Texan, but with two dominant traits—he liked a long toddy and he was a stickler for opening his court on time.

On the day he was to testify the cattleman was late, arriving at about 11 o'clock while the court opened at 9 o'clock. Judge Willis promptly fined Mose $50 for contempt of court.

The rancher waited until court adjourned and fell into step with the judge as they walked toward the hotel past a line of saloons.

"I want to explain why I was late," he said. "My wife was sick and I stayed with her." No answer from the judge. "Besides," added Hays, "I had branded a bunch of calves and separated them from their mothers, and we had some trouble with them, so I had to wait over until 3 o'clock this morning." Still no sign from the judge. "I started from the ranch and rode as hard as I could," said Hays desperately, "but when I reached the river it was up. I waited awhile and finally had to swim my horse across to get here." Stern imperturbability sat on the judge's countenance. They were passing the last saloon before they reached the hotel. "Say, Judge," remarked Mose suddenly, "how would you like a long toddy?"

The judge halted. "Mose," he said, "now you've said

something." The fine was remitted and the expression became classic in the Panhandle.

But though the courts and law of the frontier at first were not smooth in operation and worked with informality, some of which was aggravated by the inordinate propensity toward humor of the West, orderly processes gradually were brought into being, and the wild horsemen were tamed. This result was brought about inevitably from the fact that through these years the cattle country had fought a steadily retreating battle with the oncoming plow. The cowboy's natural enemy, the granger, was aided and abetted by the government. From 1860 on, Washington gave away the public lands with a hand so lavish that before the end of the century all that were fit for settlement, except for a few national forests and parks, were gone. During the period of railroad expansion railroad companies were heavily subsidized with land. All told, they received some one hundred and fifty-eight million acres, with all mineral resources and lumber under them and on them, a great and rich empire stretching in broad bands across the plains, the mountains and the deserts. The railroads did everything in their power to stimulate agricultural settlement, both to realize something from this bonus land and to promote freight and passenger traffic. Under the dual stimulation of governmental and railroad settlement activities, grew up a generation of town boosters, who held up glittering prospects of easy money to be made in booming communities. And still the government gave and gave, as long as it had any land to give away.

Last of the great areas of fertile soil taken from the cowman and turned over to the granger and town builder was the Indian Territory. Because of governmental regulations forbidding settlement, the territory remained without farms upon it for decades after settlement had closed about it on all sides. More fluid in its character than agriculture, however, the livestock industry early found lodgement in the Indian country. Trail herds passed across it in hundreds, and some of them, like those of Major Andrew Drumm, remained. The status of the cattle growers in the Territory, however, was purely that of squatters. It could not be expected that the condition would continue permanently.

North, in Kansas, was a thriving, booming city which looked upon the unsettled country hungrily. Wichita, led by Colonel Marsh M. Murdock and Dave Leahy, who published and edited the Wichita *Eagle*, was having dreams of greatness, believing that it had a destiny equal to Kansas City's, and knowing that only through the opening of a larger trade territory to the south could that destiny be achieved. Other cities in Kansas and Nebraska were enlisted in the cause, and Captain David L. Payne, a dreamy ne'er-do-well, inspired by the Wichita boosters, began organizing what were known as the Oklahoma boomers to enter the territory and settle there in spite of the government's prohibition. He entered, was driven out; entered again, was arrested; and still went back. But when his campaign, backed by the Wichita *Eagle's* editorials, was at its climax, Dave Payne died.

Here was disaster, but the movement must not stop. Congress was impressed; it must be made to feel that something like a revolt was occurring on the borders of southern Kansas. Looking about for a new leader the boomers hit upon Major Gordon W. Lillie, known as Pawnee Bill, like Buffalo Bill Cody a circus man and an old plainsman. Pawnee Bill, in Pennsylvania with his show, saw a gaudy chance for publicity, accepted the invitation, closed his tour and hurried to Wichita. He was a national figure at the time and his espousal of the cause gave the boomer movement what it needed—publicity in every paper in the country. Three thousand would-be settlers gathered in Pawnee Bill's camp on the banks of the Arkansas River, but a few days before the march was scheduled to begin, word came that the government had capitulated, and the President had signed a bill opening the land to settlement.

The cattlemen had to move out of that and other sections of the territory as one after another they were turned over to the public. The first rush for claims was in 1889. In the succeeding few years the Kiowa-Comanche and Cheyenne-Arapaho reservations were taken over, and in 1893 the last important area left to the cattlemen was turned over to the settlers. This was the Cherokee Strip. As late as 1889 it was filled with immense ranches, operating under a lease arrangement with the Cherokee Indians, and having their shipping points just north of the Kansas border. Cedarvale handled the cattle of the Osage country, with Ed Hewins the chief shipper of the district. Hunnewell and Caldwell divided the eastern part of the Strip, with W. B. Helm, George Miller, and the Wyeth Cattle Company among the important outfits using those points. At Kiowa were seen Drumm & Snyder's U brand, the P-Diamond of the Pryor brothers, Hank Creswell's Turkey Track, Tom and Billy Quinlan's Buckle, Pat Doyle's T Square, and the brands of the Comanche Pool, with the Hall Brothers, Mose Hays, Fine Ewing, Ewell & Justice, and Ishmael & Rudolph frequently seen with many others.

As cow towns, that line of southern Kansas border cities ceased existence very suddenly. From the government came an order for all cattlemen to leave the Strip, tear down all buildings, and remove all fences by 1890, in order to make ready for settlement. The last proviso was easy to obey, because the Kansas homesteaders swarmed into the country and carried away the wire and posts to fence their own farms. The cedar logs of the bunk house and outbuildings of the Drumm ranch were hauled to Caldwell and became a store building. It was a bitter season for the cattlemen. In that summer of 1890, G. W. Rourke, the Santa Fé agent at Kiowa, shipped four hundred carloads of cattle a week for three and one-half months. Not less than one hundred and thirty-five thousand cattle left Kiowa—and most of them for market. Other points along the border had business equally large. Prices were crowded down and down by the dumping of these huge numbers on the market, but there was nothing else the cattlemen could do, for to most of them there was nowhere to go. They had to sell even if it meant ruin. By fall the Strip was empty, the ranchers were gone and things were in readiness for the roaring stampede of September, 1893, known as the Cherokee Strip run.

It was right and in the line of national progress that all arable soil should be utilized for agriculture. The old-time cowman was a nomad, and stability and sober industry came to the West with the farms. But one of the greatest tragedies ever to befall a fair country occurred when some millions of Western acres that never should have been used for anything but grazing were broken with the overreach of agriculture, after the legitimate farm areas had all been taken up. It is impossible here to go into the story of the agriculturalizing of the dry plains with its disappointments and failures. "God Almighty never intended for that soil to be broken," old-timers will tell you. The plow has done atrocious damage to a once glorious landscape.

A good part of this nation should for a long time remember May 11, 1934, the date of the greatest dust storm in all history. In what had been part of the great buffalo grass empire, the "dust bowl" had developed. Before settlement, buffalo by the millions subsisted on it—it was the heart of the great southern bison range. Later, in the cattle days, hundreds of herds of longhorns and their successors, the Herefords, grazed over it. But it was plowed in man's greed—the thick, tough sod was torn up by gang plows hitched to snorting tractors, and the fine web of grass roots which held down the soil was rotted into dust. And on that day in May, 1934, a vast wind lashed the dust bowl, ripped off the topsoil and sent it in inky black clouds which towered up and up until no airplane could surmount them, to spread their malignity over the nation. Congressmen in the capitol in Washington found their teeth gritty with dust from western Kansas and the Texas Panhandle. In a smothering shroud the dust hung over cities as far east as New York and Philadelphia and two hundred miles out at sea ships reported it.

And that great dust storm was only one out of hundreds of "dusters" which before and since have occurred in the dry, sub-humid area of the high plains, once a fair expanse of grass as fine as Nature ever provided for animal life. So bad at one time became the dust that many persons in the dust bowl area suffered from pneumonia because their lungs were filled with dust. That is what the plow did to the grasslands of the high plains, and the damage is still in progress because with each dry season, more and more of the topsoil is blown away. No man knows what will be the outcome in the dry-farm areas. The dust bowl problem is mitigated temporarily in some years by rain fall, but it will be a continuing and aggravating problem for the future.

There is a boyhood memory: Working on a ranch in southwestern Kansas I rode one day over the seemingly illimitable buffalo grass prairie, hunting some lost horses that had wandered down into the wooded valley of Crooked Creek. Far to the east three strange low mounds appeared on the horizon. Curiosity caused me to ride out of my way to examine those mounds. They were three yellow-gold stacks of wheat, being put up with pitchfork and header-barge by a family of Russian Mennonites who had settled there the previous fall.

Cowboys from other ranches heard about the wheat and rode over, sometimes twenty or thirty miles to see it. Sitting in their saddles by the wheat stacks, with knees cocked over saddle horns, watching the toiling farmers with their rich brown beards full of chaff and dirt, and their low-crowned black hats almost indistinguishable as to color because of dust, the cowboys exchanged cigarettes and gibes concerning the whole idea of raising crops of any kind in that country. Had they been able to look into the future, their humor would have faded. That which they were viewing in this small field, within a few years was to root up the entire prairie—and incidentally to root the cowboys out of an occupation.

As early as 1881 General Brisbin, the author of *The Beef Bonanza*, commented on the claim made by cattlemen that the plains were unfit for civilization with "the farmers reply by plowing up a strip some ten miles wide on its [the plains country's] eastern edge each year, and raising good crops." It was true that at an average rate of about ten miles a year the plow gnawed away at the grasslands, extending the area of cultivation and cutting down the virgin pasture. But it was not until the World War came with its demand for wheat, that the truly great agricultural expansion took place. Fifty million acres were broken in two or three years. Kansas and Oklahoma and the Texas Panhandle, under the encouragement of two or three favorable years which came just at this time, began to boast of being "the breadbasket of the world." In the Northwest, also, dry farmers attempted the impossible. They did not succeed in making the land consistently productive, but they ruined millions of acres of it for grazing.

By no means, however, was the farmer solely responsible for the pathetic condition of the range country. Immense areas of the West have not known and can never know the plow. And on much of this the grass seemed, at least until recently, as irrevocably gone as if the sod had been torn by steel. In New Mexico and Arizona, level flats are too frequently gashed and marred by gaping arroyos which did not exist a few decades ago. Elsewhere the ground is bare, save for tumbleweeds, or is disastrously eroded into sand dunes. Overgrazing caused this desolation—and the stockman is to blame. On productive Missouri soils a cattleman can figure on two acres for pasture and two acres to grow feed for each animal he raises. In the Flint

Hills of Kansas and the Osage Hills of Oklahoma, six to eight acres of the rich bluestem grass are needed in good years for every head of livestock grazing. Farther west the ration may go up to fifteen acres to the animal and in some of the arid sections as many as two hundred acres of pasturage must be allowed for one cow or steer to subsist.

Reckless competition, a long series of unfavorable years, fear of the future and greed destroyed this balance. Stockmen had seen the settlers encroach upon them time after time, had seen promising grasslands taken away and turned over to the farmer, until a feeling of impermanence was upon them. Their idea became to make profits quickly before they were pushed off the range. A thousand cattle were placed where only five hundred should have grazed. For one season or two the range might stand it, but eventually even lavish Nature reached her limit and the grass was destroyed. In *The Story of the Range*, published by the Department of Agriculture, Will C. Barnes described the process:

The mountains were turned into dust heaps; the old forage plants were gnawed to the roots and so weakened that they failed to grow. Worthless weeds and annuals took their places. The willows along the streams and meadows were eaten down to walking sticks. The meadows, stripped of their green covering, dried out. The forage cover gone, the freshets tore through the meadows, leaving great gashes in the sod and soil, which cut down deeper and deeper, draining the land as successfully as any Corn Belt farmer ever drained his waterlogged fields with machinery and tiling. Then the areas grew brilliant under purple iris, a sure sign of forage-plant decadence in a mountain meadow, and their value for grazing purposes was almost gone. . . . Down in the open plains much the same process was followed.

This represents the nadir of the condition of the range —which was reached at the height of the great drouth in 1936. Yet there are signs of better days for the livestock industry in the future. No longer is the range steer a lean, rakish longhorn with a vast propensity to rove. Instead he is usually a blocky red beast, with white markings from his Hereford ancestry. He does not run like a deer at any excuse, but prefers to move sedately and with dignity, one corner at a time, although he can on occasion move with speed and activity surprising in view of the comparative shortness of leg and depth of body of a real beef animal. He requires care. Windmills and water pumps work for him and he is confined by barbed wire so that he will not wander and walk off his tallow. In recent years, moreover, the government, through the WPA, has been active in building literally tens of thousands of stock ponds to catch run-off water—and that takes a certain part at least of the gamble out of stock raising.

How much of a gamble that was—and the chance element has by no means been entirely removed from it— is shown by the figures of the great die-up during the recent drouth cycle with which was combined government subsidized slaughter. Eleven range states—Texas, Oklahoma, Kansas, Nebraska, North and South Dakota, Montana, Wyoming, Colorado, New Mexico, and Arizona —had in 1934 29,986,000 cattle, according to the figures of the United States Department of Agriculture. Four years later, in 1938, the same states had only 22,764,000 cattle—a loss of more than seven million head. Yet there was one good result even in this diminution. In selecting cattle to kill under the government program, inferior cattle usually were slaughtered. This has brought about an appreciable improvement in some herds and a general improvement throughout the West.

Today the successful cattleman no longer leaves to chance and the fortunes of the weather the fate of his herds. He lays in a supply of cotton cake to feed when winter pastures are covered with snow or ice. Frequently he grows sorghums adaptable to dry land agriculture as a source of feed, and when he can obtain the water, he irrigates alfalfa and grains. He supplies wind-breaks or open sheds where his livestock is partly protected from late winter storms. One of the big advances in stock-growing technique has been the system of herding bulls separately from the cows, so that the birth-dates of calves may be controlled. Usually the time for dropping calves starts late in February and is finished in May, so that heavy range losses, under normal weather conditions, are avoided. The modern range cow, moreover, gives enough milk to carry her calf through until grass and weaning dates early in October provide the cow an opportunity to recover from lactation and lay on fat to prepare against the succeeding winter. This is one of the direct contributions of the barbed wire fence to the modern range.

But even more important than these factors in the rejuvenation of the range is the awakening of the consciousness of the West at last to the imminent danger in which its grasslands lie. The government is taking a hand with the range conservation program, the purpose of which is to retire from cultivation all lands which have been demonstrated by experience to be unfit for growing crops. In sub-marginal areas, farmers are being encouraged to permit grass to grow back on acres plowed but unproductive. Many individual ranchmen, recognizing the peril to themselves and their entire industry, have taken an interest in private grass propagation, and an interesting experiment is under way on a widespread front, in rotation grazing, and particularly in grass seed planting.

It is interesting to know that the grand old range covering, buffalo grass, may be brought back. Until recently it was believed that the buffalo grass could not be artificially propagated—that once it was destroyed it could not again be planted. The soil conservation service, however, recently devised a machine that can harvest buffalo grass seed—a process which hitherto has baffled man—and the indications now are that the seed can be obtained in commercial quantities sufficiently large to alter the entire picture in the dust bowl. During the great drouth cycle the government sent several expensive scientific expeditions to arid and semi-arid countries overseas, such as Mongolia and the hinterlands of Turkey and Iran, seeking forage grasses that would grow in America under similar climatic conditions. The best of these imported grasses is known as the crested wheat grass, and it is being grown successfully on many ranges, especially those of the Northwest. But after the scientists had gone half-way around the world, experimentation demonstrated that after all the homely, normal buffalo grass was the most perfectly adapted to soil and conditions on the plains.

With its aggressive new program of sowing grasses where pastures have suffered, of rotation in grazing, and of restoring plowed lands to grass the range is making a courageous stand. There is reason to hope that it will beat back.

XXXVIII

WHITE FACE COWS AND LADY TOURISTS

The West has changed greatly. Economic conditions, climatic changes, the invasion of new classes of people, a growing population, all have contributed heavily to the alteration. The law has come to the range, but more than all the six-shooters and sheriff's badges, the motor car and the paved highway have tamed the cow country. You can still talk to old-timers, and there are more of them than you would imagine, who knew personally and intimately lusty figures of the old, wild frontier decades; who even played a potent part in the events of that period of drama. But to the places where Custer's troopers toiled wearily to their rendezvous with Sitting Bull and death; where Nelson Story made his long trek of travail and heroic risk up the Bozeman trail; where Billy the Kid dared the concentric rifles of the Murphy clansmen; where Charles Goodnight found Oliver Loving dying from Comanche bullets; where millions of hoofs pounded broad highways of empire across the Indian Nations—where, indeed, hundreds of epic deeds of violence and courage were performed—the automobile today purrs along smooth concrete, bearing its bored sight-seers who too often scarcely take the trouble to glance at those scenes of high achievement.

No longer is it a test of courage and physical ability to adventure out into cattle land. The West definitely has become polite. Always courteous it was, but that courtesy now is given added point by the fact that it is competitive for the all-desirable tourist trade. So, to the cow country have come society debutantes and dude ranches. Possibly the dude ranch is the most arresting change that has come over the West. No fad or transitory institution is it. Today it stands as an integral factor in the cattle range. Last year the Wyoming state department of commerce announced that in Wyoming and Southern Montana alone more than ten million dollars were invested in dude ranches, which took in for that year an income of more than three million dollars. Three million dollars is a lot of money for a country like Wyoming. It would take the raising of many cattle to produce a sum like that.

The women's magazines that circulate in penthouses and country homes each spring contain gushing articles dealing with dude ranch vacations. The Vassar or Wellesley graduate or undergraduate learns what togs to wear, which ranches are "good" and which are not, and comparisons are given in rates with ocean voyages or trips to the New England seashore. In all this discussion there is no hint or thought that the fragile and expensive modern girl or her somewhat less fragile but no less expensive mamma needs any preparation of experience, training, or more than ordinary resolution to "go Western." Indeed such speculation would be idle and unnecessary. The debutante is as safe, and quite as smart and comfortable, in a modern, well-appointed dude ranch, as she would be in a hotel. She will be guided, shepherded, watched over, and

coddled by a lean rider whose legs, greatly bowed, show
that he is no stranger to the saddle, and whose eyes on
occasion may wander wistfully to the cool, silent sweeps
of the distant mountains as he listens to feminine chatter
around him. It is a strange metamorphosis that has altered
men from top hands in the dust and whirl of the branding
corral, living the most masculine of all lives, individualists
to the core and proud of it, into escorts for young ladies
in peg-topped riding breeches, to whom a lip-stick is the
one really indispensable article. To this has come the West
in some places.

Yet the real cattle country is by no means entirely gone.
There are an infinitely greater number of serious, hard-
working cowhands than dude wranglers, and the real cattle
and horse ranches outnumber the glorified hotels known
as dude ranches. In the back country—too far from motion
picture shows and beauty shops to be really attractive to
tourists—the herds still graze and bellow, and men with
wide hats ride in the dust. In the Big Bend, in Arizona,
in up-country New Mexico, in the back areas of Wyoming
and Montana, along coastal Texas, there are still primitive
ranches. Still does the chuckwagon go out, although it is
often mounted on an automobile truck nowadays. Still is
there need for good men with ropes to work the milling
steers, and the branding iron still bites into the flanks of
bawling calves in the roundup.

Still, too, do cattlemen meet and discuss problems—in
Kansas City, Denver, Fort Worth or Chicago. And sometimes those meetings bring great reunions. In March, 1939,
for instance, there was a gathering of men at Fort Worth
the like of which may not occur again. Their average ages
were eighty-one years; without exception they had seen
the buffalo replaced by the longhorn, and that in turn by
the purebred. Their hair was white, but their leathery,
deep-lined faces, for all their years, did not lack vigor,
and their eyes gleamed with the old reckless humor of
the pristine cattleman. In that group was Ab Blocker, who

drove more than a quarter of a million cattle over the trails from the Texas ranges, who saw the markets switch from Abilene, Ellsworth, Newton and Wichita, to Dodge City, Ogallala, Cheyenne and Miles City, and visited all of them with his herds. On that day he might have claimed that of all men then alive he had ridden the most miles on horseback and looked down on the backs of the most horned cattle.

A second in the group was Ellison Carroll, who was the world's champion roper in the '80's, when a rope was not merely something to use in tricks of sleight of hand, but an implement with which to make a living. So far as is known he still holds the world's record for roping and tying a carload of twenty-two steers in an average of twenty-two and one-fifth seconds each.

John Arnot was another, a sturdy son of Aberdeenshire, Scotland, who was one of the young Britons who came to this country in the '80's, fought their way through the black cattle depression and remained to add their courage, foresight and enterprise to the rebuilding of the cattle domain. He is one of the last of the original group which included Murdo MacKenzie, John Clay, Moreton Frewen, John Adair, and W. J. Tod, and his whole life was devoted to the developing of the L X and other great ranches.

There also was Bob Beverley, who went to work on the Pecos when he was only eleven years old, and has served as a cowhand, wagon boss, and foreman almost continuously since—a great rider, roper, and raconteur.

And in that list was Captain John Hughes of the Texas rangers, the man who broke up the border rustlers and arrested the bloody outlaw Chacon. He could tell stories if he chose of tense moments when the spurt of a six-shooter's flame lit up the darkness of a rainy night, when men gasped and died, when a campfire was a danger spot, and when all men took care to ride with nobody behind their backs.

Within their personal experiences these men combined a

knowledge of most of the history of the cattle range since the great trail-driving days began. And that in itself shows how very recent is the history of those mighty years, whose excitement makes them seem remote and legendary.

By no means has the problem of the cattle rustler, that pirate of the cow country, been abated. Today the stock thief substitutes speedy rubber-tired trucks, paved roads, flashlights, and dishonest butchershop "fences" for fast horses, six-shooters, and running irons. Yet in the essential he remains the same abhorrent, despised and feared threat on the range as of old.

Cattle inspectors, one day in 1938, discovered the hides of four hundred cattle which had been tossed deep into the yawning gulf of Cucharas canyon, near Walsenburg, Colorado. The brands and ear-marks of these animals all had been cut out so the owners could not identify them. At about the same time, a rustler, captured and convicted at Greeley, Colorado, made a written confession that he had stolen fifty-eight cattle in fifteen months. And this was the swift and efficient method of operation he described, which is fairly typical of the habits of all the rubber-tired rustlers:

The rustler had a truck built on the lines of a moving van—in fact, a sign on the sides in many cases actually advertises that it is a furniture transportation vehicle. Selecting a dark night he drove to the pasture of a ranch whose lands happened to lie convenient to a highway. There the wire fence was cut with a few snips of the wire-cutters, and the truck driven boldly into the pasture, into some low place such as a ravine where any lights lit might be protected from observation. Inside the high-walled van were the horses of the gang, usually two or three men to a truck. The horses were unloaded and mounted, and two or three likely steers rounded up and driven to the gully, where a rifleman, using a Maxim silencer on his gun, dispatched them. All the rustlers were expert butchers. Within a short time the animals were skinned, cut up, and the meat placed in the truck. There was room also for the horses; the rear endgate was closed, the engine roared, and the pseudo moving van rolled away through its self-created gap in the fence, leaving only the hide and offal for the owner's rueful contemplation the following day. By dawn, that truck with its stolen beef might be two hundred miles away and in another state, with the meat in the cooler of some unscrupulous butcher.

The stockmen have made repeated efforts to enlist the aid of the federal government in fighting the rubber-tired rustler, and in the 1939 session of congress a bill was actually passed making it a federal offense to transport stolen cattle across state lines, on the same theory that the Dyer act makes illegal the transportation across state lines of stolen automobiles. President Roosevelt, however, vetoed the bill on the ground that already there was too great a tendency to make policemen out of government officials, and pointed out that, carried to its logical conclusions, such a law would eventually set Uncle Sam to trailing chicken thieves.

So the cattle associations have experienced a new resurgence of life. The cattlemen are used to doing without aid from the government and they are good at standing on their own feet. In Kansas a branding law has just been enacted after years in which the branding of cattle had been allowed almost to lapse. All through the Western range country it is common to see, posted on telegraph poles, pine trees, or public buildings, signs that read like this.

$1,000 REWARD FOR INFORMATION LEADING TO ARREST AND CONVICTION OF ANY PERSON STEALING LIVE STOCK OF ANY MEMBER OF COUNTY LIVE STOCK ASSOCIATION

Those signs indicate the revival of small local associations of cowmen which once flourished throughout the West. They also are declarations of a new war by the

aroused cattle industry against the most modern form of
range outlawry. On many leading highways in the West,
brand inspectors are stationed, and legal papers must be
shown to inspectors who question the ownership of any
beef or animals found. The radio also has been called in
to help capture rustlers. Vigilantes, as stream-lined as the
stock thieves themselves, patrol back roads in speedy automobiles. New laws, with more teeth in them, have been
passed in many states, aimed at rustling. It is a long fight,
but the cowmen are in the saddle again, and it is not beyond conception that there may be some vigorous days
ahead.

The old trail-driving days are gone, probably forever,
and the great names connected with them are no longer
active—with one notable exception. Down near Jermyn,
Texas, the estate of Oliver Loving still operates the Turkey Track ranch and sells many cattle yearly. With this
exception the new generation of ranchers sounds strange
to the ears. There are no Ike Pryors, Charles Goodnights,
John Chisums, John Iliffs, or Shanghai Pierces riding the
trails these days.

Yet even in the great cattle migrations there is a modern
counterpart. In Southeastern Kansas and Northeastern
Oklahoma lies the famous bluestem grass country. Over
several counties in both states it spreads—the Flint Hills
in Kansas and the Osage Hills in Oklahoma. The tall, rich,
jointed bluestem grass comes closer than any grass grown
on the globe to giving cattle a finish akin to that produced
by grain. And each spring occurs an amazing hegira to
that section. From Texas, New Mexico, Arizona, and even
from deep in Old Mexico, long trains of loaded cattle cars
puff toward the bluestem pastures.

In 1929 the total number grazed in the bluestem was
three hundred thousand in Kansas and two hundred thousand in Oklahoma—half a million all told—and the average
annual number is not far below that. With the single ex-

ception of 1871, when more than six hundred thousand cattle were driven up the trails out of Texas, it is doubtful if the trail drivers ever in one year brought more than half a million cattle northward. By the standards, therefore, even of the great trail-driving days, the annual migration to the bluestem area is impressive.

And along this line something happened recently in New Mexico which would have delighted Richard King, Shanghai Pierce, Jim Blocker, Ike Pryor, and others of the great trail-driving generation. The United States government, through the Civilian Conservation Corps, began the building of a stock trail from the upland grazing country of New Mexico and Eastern Arizona, to Magdalena, New Mexico, terminus of an inland branch of the Santa Fé. It is a veritable dream of a trail, seventy-five miles long and a quarter of a mile to four miles wide, with controlled grazing and modern watering facilities. What would the old cattleman who fought for a national stock trail from Texas north have done with something like that!

Cowboys still sing to their herds in the wild remote areas of the cattle country. Still, too, there is a depth of sadness in their melodies, a tinge of melancholy fatalism:

> They say there will be a great roundup
> Where the cowboys like cattle must stand,
> To be cut by the riders of Judgment,
> Who are posted and know every brand.
>
> Perhaps there will be a stray cowboy,
> Unbranded by anyone nigh,
> Who'll be cut by the riders of Judgment
> And shipped to the sweet by and by.
>
> I don't see why there's so many
> To be lost at that great final sale,
> Who might have been rich and had plenty
> Had they known of that great final sale.

In most ranches, however, the night herd is pretty much a thing of the past. The great ranch properties which saw

their golden era in the early '80's, then withered in the
face of the blizzard years and the panic years, have been
to a large degree broken into smaller, more compact and
more efficient holdings. Yet it is possible to operate on a
scale too small, and the range country has learned that
efficiency will not permit operation, strictly in cattle, with
fewer than five hundred head. Modern methods, also, have
halted the break-up of the remaining big ranches. The
immense King ranch of southern Texas recently has increased its holdings by new purchases. The Matador Land
& Cattle Company still maintains one of the best-run establishments, large or small, in the West, and it numbers its
acres by the hundreds of thousands. There are dozens of
other large ranch properties still in existence in the West
and apparently they are going to stay.

Good bulls, good fences, good horses, permanent headquarters with well-established central offices and comfortable accommodations for the men, good chutes, corrals
and water development—these are the hall-marks of the
up-to-date modern ranch. It took time to make the change
which changing conditions in the range country necessi-
tated, but by and large the cowman has made that change
surprisingly well.

Of important bearing on this is the fact that the old
enmity between range and grange has disappeared. Fairly
well at last the differentiation between tillable and non-
tillable soil has been established, and the cattleman has a
fair notion that he will be permitted to remain where he is.
The result is a growing partnership between corn belt and
cattle range, where only a feud existed before. The State
of Texas and other range states have gone into the mammoth task of eradicating the cattle tick which formerly
spread disease. Southward, ever southward, the tick deadline has been pushed, until today it exists only in extreme
southern Texas and it is being shoved back still farther
yearly. Last year Texas alone dipped twenty million animals—chiefly sheep and cattle—to safeguard against tick,

nor is there any diminution in the war expected in the future.

Under such conditions farmers in Missouri, Iowa and Illinois are delighted to obtain steers from the grassy ranges to feed with corn and sell to the packer. But a still further development is being forecast. During the trail-driving era, some of the important cattle companies, such as the Matador, X I T, X L, and others, reared cattle on the Southern plains, then sent them up to the Northern plains to mature and fatten. There are signs that a similar arrangement may obtain in partnership with the corn belt—cattle raised on the plains shipped to the owner's own property in the corn belt, fed on the owner's corn, and sold directly by the owner to the packer, thus eliminating duplications in transportation costs and middleman's profits.

The cow country and the corn belt in close alliance? It sounds somehow incongruous, the mating of opposites, the abatement of the age-long, epic feud of Cain and Abel. And it leaves also a feeling that this may mean the final step in the transmutation of the West into a land of safe and sure progression and dull conformity. Yet that this should occur I think doubtful. The West has accepted modernity to a great degree. The motor car narrows its horizons and even the windmill has been impressed into service to charge storage batteries so that the radio may be listened to in remote cabins.

Not long ago I watched four cowboys riding slowly toward me across the sagebrush flats of New Mexico. From a distance I took in the details of their costume— wide hat to high-heeled boots they were the typical riders of cattle land. Their saddles were familiar in appearance, the very way in which their ponies jogged, scuffing through the dust, was appreciated by my inner eye, because it was a thing well understood and well remembered. But something about the distant riders bothered me—some detail on which I could not at first place my finger, but

which struck a jarring note. Then—as they came closer—
I saw what it was. Every cowboy in the group wore sun
glasses of the type popularized by Hollywood and now a
part of the summer costume of every society girl.

The first uneasy feeling was that this was a thing not
quite right. But after that preliminary twinge I suddenly
knew that it was right. Dark glasses—of course! If there
is one person who really needs them, it is your cowboy,
riding all day and day after day, under the merciless desert
sun. And the incident somehow typifies the situation in
cattle land today. Innovations are adopted, but the adoption is characterized by common sense. In the essentials of
space and open-heartedness and the high value placed on
individual independence, the range country remains much
the same. Towns are still far apart. The barbed wire is
there, but its web of spiked steel is not of such nature that
it obstructs the scenery—the magnificent scenery of mountains and deserts and plains. And the sunset continues to
be unmatched in glory.

SOME BOOKS TO READ

Formal bibliographies are tedious and, except for the serious researcher, practically valueless. The busy man or woman, interested in a subject, wishes to know the best places to go for added knowledge, not all the places. Therefore, although the sources I have used in this volume are many, including not only books, but newspaper files, manuscripts, and personal narratives, I will suggest here a comparatively few books only, and these, for interest and factual content, I can unreservedly endorse.

For the early history of the cattle industry, the starting place for a reader is Joseph G. McCoy's *Historic Sketches of the Cattle Trade*, written by the man who more than any other was responsible for the inception of the great trail driving era. Going still farther back into early Texas history is the monumental *Historical and Biographical Record of the Cattle Industry* by James Cox. And anyone who tries to read of the trail-driving period without including *The Trail Drivers of Texas* by J. Marvin Hunter, compiler, and George W. Saunders, president of the Old Time Trail Drivers' Association, who supervised it, will have overlooked one of the richest of all sources of material. Of this period, including some exciting Indian warfare, James Cook also writes in *Fifty Years on the Old Frontier*.

There are some modern classics on the cattle industry. Among these none is superior to J. Frank Dobie's *A Vaquero of the Brush Country*, and Mr. Dobie has written also an invaluable study entitled *The First Cattle in Texas and the Southwest, Progenitors of the Longhorns*, which traces the development of the early cattle breeds and the early cattle ranches. *Cattle*, by William McLeod Raine and Will C. Barnes, is a vivid general view of the range

417

country and its history, and Struthers Burt's *Powder River* captures the spirit and color of the cow domain as have few writings on the subject. In this category belong also *The Cowboy*, by Philip Ashton Rollins, and *The Story of the Cowboy*, by Emerson Hough.

For the serious student of economic or social trends, I suggest *The Range Cattle Industry* by Edward Everett Dale, *The Cattleman's Frontier* by Louis Pelzer, and William Curry Holden's interesting *Alkali Trails*. The classic personal reminiscence of a big cattle operator, giving the financial problems of a large rancher, is John Clay's *My Life on the Range*, and the collateral viewpoint of the packer can be obtained interestingly from *The Yankee of the Yards, the Biography of Gustavus Franklin Swift*, by his son Louis F. Swift and Arthur Van Vlissingen, Jr.

As to the era of the cattle trail towns, I know of no better or abler work than Floyd B. Streeter's book *Prairie Trails and Cow Towns*. A valuable work also is William McLeod Raine's *Famous Sheriffs and Western Outlaws*, as is Emerson Hough's *The Story of the Outlaw*. More difficult to obtain but invaluable is *Dodge City, the Cowboy Capital*, by Robert M. Wright, a sprightly, though disjointed account of frontier Dodge. For a settler's view of the Abilene difficulties, see Stuart Henry's *Conquering Our Great American Plains*.

A unique biography of a unique character is *Charles Goodnight* by J. Evetts Haley. In this story of the cow country's most remarkable figure, almost the whole panorama of the cattle range is contained.

For the period of the range wars, there is a large and vigorous literature. Naming only a few, I can suggest: for the Lincoln County war, *The Saga of Billy the Kid*, by Walter Noble Burns; for the Cochise County war, *Wyatt Earp, Frontier Marshal*, by Stuart Lake, and Billy Breakenridge's *Helldorado;* for the Tonto Basin war, there is no book more complete or authoritative than *Arizona's Dark and Bloody Ground* by Earle R. Forrest. To obtain a view of the Northern aspects of the sheep and cattle war and also for a simple yet fervent close-up account of the Johnson County war, read A. J. Mokler's *History of Natrona County* (Wyoming). *Malcolm Campbell, Sheriff,* by R. B. Davis, is a very vivid and complete history of the Johnson County war.

To the reader interested in the side of the nester and pioneer farmer in the settlement of the West, Everett Dick's *The Sod House Frontier* is an able and interesting volume, and for him who wishes to read of the strange and tragic history of the buffalo in this country I can cite nothing better than M. S. Garretson's *The American Buffalo.*

There is a vast and voluminous literature on the Indian wars. One of the best on the Kiowa-Comanche troubles is *Carbine and Lance* by Captain W. S. Nye. The fighting with the Sioux is ably described in *Sitting Bull* by Stanley Vestal, and my own *Death on the Prairie* and *Death in the Desert* give a general view of the Indian fighting in the West from the time of settlement on.

By no means does this listing even begin to give a bibliography of the West. But it will serve to give a substantial background of knowledge, and it is to be hoped that it will stimulate the reader to go more deeply into the rich history of the cattle country.

INDEX

Aber, Doff, 272

Abilene, rise and fall of, 132 ff., 144, 169, 170; kind of settlers, 156, 157; cowboy antics in, 157-162; law enforcement in, 163-169

Acevedo, Gaspar de Zuniga y, 21

Adair, John G., 279-280, 408

"Alamo," Abilene, 159, 167

"Alamo," Newton, 180

Alexander, Milton, sheep killer, 371

Allemand, Joe, 371

Allen, Jack, 196

American Bison Society, 218

American National Live Stock Association, 252

Amusements, cowboy, 255-264

Andalusian cattle, introduction of, 14, 16

Anderson, Mrs. Annie, quoted, 199

Anderson, Bill, 192

Anderson-McCluskie episode, 183-185

Angus, Sheriff Red, 390, 391

Angus cattle, 318

Anson, William, 267

Applejack saloon, Abilene, 159

Arizona, cattle introduced into, 24; ranges in, 151; herds driven to, 316; rustlers' war in, 356 ff.

Armour, P. D., pork packer, 73, 74, 77, 95

Arnot, Jack, 408

Askew, Kirk, saddlery firm, 236

Audubon, 138

Austin, Stephen A., 36, 42, 56

Axtell, Gov. Samuel B., 345

Aztec Land & Cattle Company, 372

Baker, Col. E. M., 217

Baker, Frank, in Murphy-Mc-Sween feud, 339, 344

Barbed-wire fencing, 205, 286 ff., 316

Bar C ranch, 282

Barnes, Will C., quoted on grass shortage, 401

Barry, Buck, 36

Barton, D. W. ("Doc"), trail maker, 193-194; loss in blizzard, 307

Basset, Charlie, 196

Baxter Springs, 89-90, 94, 130, 156, 157

Bean, Judge Roy, 393

Beard, John ("Red"), 191

Beard-Lowe episode, 191-193

Beckwith, Bob, 351

Beef, demand for, 57, 221

Beeson, Merritt, quoted, 198

Behan, Sheriff Johnny, 360, 364, 365

Bell, J. W., 353

Belle Fourche Cattle Company, 300

Bell ranch, 381

Berry & Boice, 282

Beverley, Bob, 408

"Big Hank," encounter with Smith, 163-164

"Billy the Kid." See Bonney

Birdsong, Mrs. Neva, 48

Bison, American, 17

Black Beaver, scout, 138

"Black cattle," 36

"Black Friday," 219

Blevans, Andy ("Cooper"), 374, 376, 387

Blevans, Charles, 374, 378

421

Sun glasses, introduction of, 415
Swan Land & Cattle Company, 253, 300, 310
Sweet, Jim, 186, 191
Swenson, S. M., 314
Swift, G. F., meat packer, 73, 74-78
Swift, Louis F., cited, 73
Swilling, Hank, 365
Swimming steers, 68-69, 119 ff.

Tafoya, José Pieda, 215
Tallow, development of market for, 57, 58
T-Anchor ranch, 282
Tankersley, R. F., 282
T A ranch, 390-391, 392
Tascosa, 156, 281
Taylor, Gen. Zachary, army of, attacked by bull, 37-38
Ten Sleep war, 370-371
Tewksbury, Ed, 372, 379
Tewksbury, Jim, 372, 375, 379
Tewksbury, John, Jr., 372, 376
Tewksbury, John D., Sr., 372
Tewksbury-Graham feud, 371-380
Texan-Sante Fé expedition, 315
Texas, cattle introduced into, 21, 22, 33 ff.; Indian warfare, 40 ff.; effect of annexation, 51; land grants, 53; number of cattle in, 54, 56; cattle industry, 56; hide and tallow trade, 56-58; trail driving, 58-61, 83 ff.; in Civil War times, 63 ff.; return of ex-soldiers to, 78-82; influence of, on cattle industry, 230 ff.; exodus from Panhandle, 315
Texas and Southwestern Cattle Raisers Association, 251-252
"Texas fever," 85-87, 140. See Ticks
Texas rangers, 44-46, 47, 315, 408; war with Comanches, 48-52; Colt revolvers adopted by, 314
Texas Trail Drivers' Association, 225
Thode, Earl, 272

Thompson, Ben, 158, 166 ff.; arrests himself, 173; gambling brawl and death, 174-177
Thompson, Billy, 166, 172; shoots Whitney, 174 ff.
Three 7's ranch, 282
Throwing steers, 240-241
Ticks, disease carried by, 85-87, 140, 283-285, 413-414
Tilghman, Bill, 196
Tin Can Alley, Dodge City, 201
Tisdale, Bob, 385
Tisdale, J. A., 385
Tod, W. J., 310, 408
Tombstone, Arizona, 356
Tongue River, 210
Tonto Basin war, 371-380
Tourist trade, 406
Towns, cattle trading, 156-162, 397
Trail branding, 113
Trail driving, to New Orleans, 58; to Ohio, 58; to New York, 58-59; to California, 59-60; Saunder's adventure in, 64-67; to Mississippi, 66-67; Snyder's swimming steers in, 68-69; to Confederate Army, 68-69; end of, in Texas, 69; through Indian Territory, 83-85; local opposition to, 85-87; Jayhawkers' war on, 87-89; to Wyoming, 96 ff., 152; Goodnight's expedition, 100 ff.; methods, 110-118; profits, 111; general details of, 121 ff.; cowboy life during, 153 ff.; to the Northwest, 219-227; decline, 225-228
Trailing, herd, 110-111. See Trail driving
Trails: Pecos, 101 ff., 147 ff., 205, 222; Chisholm, 134, 137-139, 162, 224, 225; Western, 224, 225, 227; present-day, 412
Transportation. See Railroads
Tucker, Tom, 375
Tumlinson, Captain, 44
Tunstall, J. H., 337 ff.
Turkey Track ranch, 282, 411
Turner, Marion, 347

Union Stockyards, Chicago, 73, 76
United States, first cattle to enter, 18 ff.
U ranch, 287

Vaca, Cabeza de, 21
Valle de Las Lagrimas, 211
Van Dorn, Maj. Earl, 51
Vaqueros, life of the, 26-31
Villa, Pancho, 250
Villalobos, Gregorio de, cattle shipped by, 14, 16-17

Waggoner, Dan, 225, 282
Waggoner, Tom, 386
Wagoner, Jack, 197
Walker, Alfred, 197
Walker, Bill, 386-387   [314
Walker, Capt. Sam, gun inventor,
Wallace, Big Foot, 44
Wallace, Gen. Lew, 351
Ward, Seth E., 143
Water, modes of supply, 292-294
Water rights, 30-32, 229, 316; Iliff's, 207
Wayte, Fred, in Murphy-Mc-Sween feud, 344
Western trail, 224, 225, 227
Wheat, introduction of, 399-400
Wheeler, Col. O. W., caravan led by, 134-135
White, Fred, 361, 362
Whitney, Chauncey B., 173, 174, 178
Whitney, Eli, as gun maker, 315
Wichita, Kansas, 156; cattle boom days, 190-193; development of, 396
Widermann, Robert A., 339, 345
Wild cattle, attacks by, 36-39; taming, 97-110

Wilkerson, Jim, 185
Williams, Col. Len, 46
Williams, Mike, 168
Willis, Judge, 394-395
Willoughby Bros. & Drumm, 203
Wilson, Billy, 378
Wilson, John P., 343
Wilson, "One-armed Bill," 148, 150
Wilson, Tom, cowboy, 119
Windmills, introduction of, 293
Wire-cutters' wars, 321 ff.
Wire fencing. See Barbed wire
Wolcott, Maj. Frank E., 385, 390-391
Wolf Mountain, battle of, 217
Women in the cattle country, 259-262
Woods, Eddie, 272
Wootton, Uncle Dick, 105, 151-152
WPA activities, 402
Worsham Cattle Company, 310
Wright, Bob, 197
Wright, Robert L., cited, 194
Wyeth Cattle Company, 397
Wyoming, 143, 207, 216, 316; Indian raids in, 221; herds brought to, 222; influence of, on cattle industry, 231; sheep war in, 370 ff.
Wyoming Cattle Ranche Company, 300
"Wyoming Frank," encounter with Smith, 164-165
Wyoming Stock Growers' Association, 251
Wyoming Wool Growers' Association, 371

XI ranch, 238
XIT ranch, 227, 282; barbed wire used on, 289-292

WASHINGTON

MONTAN

OREGON

COLUMBIA PLATEAU

CASCADE MOUNTAINS

NORTHERN ROCKIES

IDAHO

Yellows

Snake R.

BIG HORN BASIN

WYOMIN

CONTINENTAL

GREAT BASIN

ROUTE TO CALIFORNIA

GREAT SALT LAKE

SIERRA

CARSON CITY

STOCKTON

SAN FRANCISCO

NEVADA

DESERT

UTAH

Green R.

Grand R.

CO

NEVADA

CALIFORNIA

DESERT

LOS ANGELES

Colorado R.

ROUTE TO CALIFORNIA

ARIZONA

NEW M

ALBUQ

HISTORICAL COLLECTION
✤ OF ✤
MILTON LANYON

Gila R.

YUMA CALIFORNIA ARIZONA TRAIL

R
LIN
CR
GE

A
Map of the Cattle
Country showing
the Chief Herd Routes